Y0-ACG-091

12-14-72

UNITS OF
MEASUREMENT

Units of Measurement

An encyclopaedic Dictionary of Units
both scientific and popular
and the quantities they measure

Stephen Dresner

Hastings House, Publishers
New York 10016

Published in Great Britain by
HARVEY MILLER & MEDCALF LTD
Aylesbury, England

SBN 85602 002 8

First Published in the U.S.A., 1972 by
HASTINGS HOUSE, PUBLISHERS
10 East 40th Street
New York 10016

SBN 8038 7496 0

Library of Congress Catalog Card Number 78–187316

Book Designed by George W. Hynes

Printed in England by Hazell Watson & Viney Ltd
Aylesbury, Bucks

Preface

Eight hundred and fifty years ago the size of a roll of cloth purchased in this country was measured—albeit indirectly—against the length of King Henry I's arm. In 1791 the stated height of a French horse depended on the distance between Dunkirk and Barcelona. For a little less than a century the distance of a star has been popularly given in terms of the speed of a ray of light. The historical development of different methods of measurement has resulted in a wide variety of units. Although often the original meanings were independent, most units of a given quantity are today formally defined in such a way that they are directly related to one another. The main purpose of this book is to provide formal definitions and explanations; this will enable the user, who may be unfamiliar with a particular unit, to connect it with one he understands.

At the present time the United Kingdom is embarked on the early stages of 'metrication'. It is to be expected that in due course all measurements will be made in metric units, and in particular in the units of the International System (SI). Most other countries that are not already fully metric are changing in a similar fashion. The majority of measurements in science are already carried out in this system. It appears logical, therefore, to give the size of each unit in terms of that of the SI unit.

Because the full understanding of a unit depends on many other factors, and in particular on the quantity that it measures, the book includes an extensive section on the most important quantities. A series of appendices provides a large amount of related information.

It is a great pleasure to record my thanks to two people in particular. Tony Bourne has investigated the meanings and uses of several less well-known units, particularly in the field of biological science; and my wife has devoted much time to routine but essential checks and counter-checks.

STEPHEN DRESNER
London: April 1971

Table of Contents

List of Tables

PART 1: UNITS (Tables unnumbered)

PART 2: QUANTITIES (Tables unnumbered)

PART 3: APPENDICES

Introduction

PART 1: UNITS

1 Name of the unit

1–1 Units, the names of which consist of more than one word, are entered with the most important word first. This is usually (but not always) the only substantive in the name.

1–2 Where the name of a unit can be spelled in more than one way the commonest or preferred spelling is used. A note draws attention to the alternatives.

1–3 Some names are given to more than one unit. The names are separated either by a distinguishing comment in parentheses immediately following the name [eg second (of arc), second (of time)], or by numbering the usages.

1–4 Some units have more than one name, and are entered under each. A note draws attention to the other names, and it is so worded that the preferred name is indicated.

2 Unit symbol

2–1 If there is only one unit symbol it is placed immediately after the name of the unit. If there are two or more unit symbols they are entered under a special subheading, and an order of preference is indicated.

2–2 An unqualified unit symbol conforms to a Recommendation of the International Organisation for Standardisation (ISO). (BS) following the unit symbol means that it is recommended by the relevant British Standard, but does not appear in an ISO Recommendation. (O) following the unit symbol means that it is in common use, but has not received formal recognition.

2–3 Alternative unit symbols are sometimes required because otherwise two units with same symbol may be used together.

3 Quantity measured by the unit

3–1 This is entered under its own subheading. In those rare cases where the quantity is not included in Part 2 (Quantities) it is explained.

3–2 If the quantity measured can have several different names, the basic name used as an entry in Part 2 (Quantities) is given: the other names are listed under this entry.

3–3 It should be noted that the term 'volume' normally refers to a length cubed [eg cubic metre, cubic foot], and 'capacity' to other measures [eg litre, gallon].

4 System to which the unit belongs

4–1 This is entered in parentheses after the quantity measured by the unit. The systems are summarised in table 1 of Appendix 1.

4–2 Some units are characterised by the following descriptions:
- (a) metric — derived from the SI and CGS systems by a power of ten;
- (b) metric-derived — metric, but derived by a value that is not a power of ten;
- (c) imperial — derived from the FPS (or a similar) system;
- (d) all — belonging to both metric and imperial systems;
- (e) arbitrary — defined (often as a result of an experimental investigation) without reference to an established unit system;
- (f) none — the unit measures a dimensionless quantity.

5 Definition

5–1 A verbal definition is given in every case except where the unit is a multiple or submultiple of a basic metric or imperial unit.

5–2 The formal French definitions are given of the fundamental SI units, followed (in parentheses) by literal translations that, where necessary, have been modified to bring the definitions completely up to date.

5–3 A defining equation is given in every case except where the verbal definition does not allow it.

5–4 The object of the defining equation of a metric or metric-derived unit is the unit size in terms of the corresponding SI (International System of Units) unit. In the former case it is an exact conversion; in the second case it may or may not be.

5–5 Most other defining equations have more than one object. The first is the unit size in terms of the basic unit of the same system, and is therefore exact. It should be regarded as the formal definition. The last object is the unit size in terms of the corresponding SI unit, and may or may not be an exact conversion.

5–6 The defining equation is immediately followed by a reciprocal defining equation, which gives the size of the corresponding SI unit in terms of the unit concerned. In many cases this is not an exact conversion.

5–7 The sign $=$ represents an exact equivalent. The sign $\approx$ means that the object of the equation is correct to the number of significant figures quoted.

5–8 The rule for the number of significant figures is as follows:
  (a) When the conversion is exact (ie it may be expressed by a terminating decimal), all the significant figures are included.
  (b) When the conversion is inexact (ie the decimal value involved is non-terminating), six significant figures are given.
  (c) When the value is based on a quantity experimentally determined to x significant figures, the conversion value is also given to x significant figures.

5–9 British Standards often provide full conversion tables [eg metre to inch, foot, mile, etc; and reciprocal values]. Since the exclusive use of SI units is becoming very common in both scientific and non-scientific measurements, it has been felt necessary to give conversion values to and from the SI only.

5–10 Extensive tables of unit relationships for the more common quantities are given in Appendix 5 [eg 16 ounces $=$ 1 pound; 14 pounds $=$ 1 stone; etc].

5–11 If the size of a unit as used in the United Kingdom (UK) differs from the size of the corresponding unit used in the United States (US), the name and unit symbols are modified by the addition of the letters UK and US.

6 Notes

6–1 These are numbered and entered under the final subheading.

6–2 The comment 'popularly' means that the usage is common, although it is not found in specialised or scientific works.

6–3 The comment that a unit 'has been called' by another name suggests that the alternative name is no longer used.

7 Units not included in Part 1 (Units)

7–1 Metric multiples and submultiples of units (see Appendix 3, section 2). The exceptions to this rule are:
    (a) the kilogramme and its direct derivatives, since the kilogramme is a fundamental SI unit;
    (b) multiples and submultiples of units generally not employed in the basic form.

7–2 Combination units [eg metre per second]. There are a few special exceptions, where the combination form is often thought of as a unit in its own right [eg kilowatt-hour].

7–3 Obsolete and old-fashioned units: see Appendix 4. Also included in this appendix are units used only for certain commodities [eg the board foot, for timber].

7–4 Foreign, archaic (eg Roman and Greek) and biblical units.

8 Cross references

8–1 Units mentioned in the notes to other units are also entered in their correct alphabetical position.

8–2 Units with names that consist of two or more words are only entered under one word of the name (see 1–1), and not under each word.

9 US variations

Although formerly the sizes of many imperial units were given definitions different in the United States from those employed in the United Kingdom, these differences have now been eliminated in all cases except the following:
    (a) Units of capacity. In the UK the gallon is defined somewhat arbitrarily by the 1963 Weights and Measures Act. In the US the gallon is defined in terms of the cubic inch.
    (b) The hundredweight and the ton have different formal definitions.
Where these variations exist, they are clearly pointed out (see 5–11).

PART 2: QUANTITIES

10 Name of the quantity

10–1 Quantities, the names of which consist of more than one word, are entered with the most important word first. This is usually (but not always) the only substantive in the name.

10–2 Some names are given to more than one quantity. The names are separated by numbering the usages.

10–3 Some quantities have more than one name, and are entered under each. A note draws attention to the other names, and is so worded that the preferred name is indicated.

11 Symbol

11–1 If there is only one symbol it is placed immediately after the name of the quantity. If there are two or more symbols they are entered under a special subheading, and an order of preference is indicated.

11–2 An unqualified symbol conforms to a Recommendation of the ISO. (BS) following the symbol means that it is recommended by the relevant British Standard, but does not appear in an ISO Recommendation. (O) following the symbol means that it is in common use, but has not received formal recognition.

11–3 Alternative symbols are sometimes required because otherwise two quantities with the same symbol might occur in the same equation.

12 Definition

A verbal definition is given, followed by a defining equation. Sometimes an alternative verbal definition and defining equation are also given.

13 Unit of measurement

This is entered first under the subheading 'unit'. For dimensional quantities, only the SI unit is given, except where its use has not been approved; then the commonly-used unit is given. For dimensionless quantities which possess units the recommended unit is given, sometimes with an alternative. A full explanation of the unit will be found under the appropriate entry in Part 1 (Units).

14 Dimensional forms

These are entered directly after the unit. The systems employed are discussed in Appendix 2, sections 4 to 7.

15 Type of quantity

This is entered in parentheses after the dimensional form. The descriptive terms used are:
- (a) scalar (ie a quantity possessing size only);
- (b) vector (ie a quantity possessing size and characterised by a direction of action).

16 Notes

16–1 These are numbered and entered under the final subheading.

16–2 In the case of an electrical quantity in which the rationalised and un-rationalised sizes are not equal a special note, marked (U/R), is included. In it, the first equation gives the connexion between the two quantities; the second equation gives the definition of the unrationalised quantity (which is distin-guished by carrying a star) that corresponds to the formal definition of the rationalised quantity given previously. The absence of such a note auto-matically implies that the rationalised and unrationalised quantities are identical.

17 Cross references

17–1 Quantities mentioned in the notes to other quantities are also entered in their correct alphabetical positions.

17–2 Quantities with names that consist of two or more words are only entered under one word of the name (see 10–1), and not under each word.

18 Quantities included in Part 2

The quantities chosen for inclusion by no means constitute a complete collec-tion. The main reason for including a quantity was the feeling that it is 'basic', ie it is likely to be used in various different circumstances. Its nature should be sufficiently clear so as to be useful to many investigators, and not just to those working in a highly specialised field.

GENERAL NOTES

19 Abbreviations employed

(BS)   Recommended by the relevant British Standard, but does not appear in an ISO Recommendation

CGPM   Conférence Générale des Poids et Mésures (General Conference on Weights and Measures)

CGS   Centimetre – gramme – second (system)

CIPM   Comité International des Poids et Mésures (International Committee on Weights and Measures)

FPS   Foot – pound – second (system)

IEC   International Electrotechnical Commission

ISO   International Organisation for Standardisation

(o)   In common use, but has not received formal recognition

SI   Système international (international system of units)

UK   United Kingdom (generally indicating a variation when compared with US)

US   United States (generally indicating a variation when compared with UK)

=   Equals, ie is an exact equivalent

≈   Equals, to as many significant figures as quoted in the object of the equation

20 Purpose of the Index

Included in Part 1 are a number of units which measure very specialised quantities not given in Part 2 (Quantities). Reference to the quantities can be found through the Index. Also, there are many ideas and comments scattered throughout the book (especially among the appendices) that can quickly be located using the Index.

PART 1
UNITS OF MEASUREMENT

AB-...

The names of units of unrationalised electrical and magnetic quantities in the CGS-emu system are obtained by prefixing the corresponding SI unit names by ab- (representing the words 'absolute electromagnetic'). These units are not the same as absolute units. The following table contains single-word units only.

CGS-emu unit	unit symbol	quantity measured	corresponding CGSm unit of equal size	corresponding CGS-emu mechanical unit	size of one CGS-emu unit in SI units
abampère	abA	current	Bi	$\mathrm{dyn}^{\frac{1}{2}}$	10 ampères
abcoulomb	abC	charge	Bi s	$\mathrm{dyn}^{\frac{1}{2}}$ s	10 coulombs
abfarad	abF	capacitance	$Bi^2\ s^2/erg$	s^2/cm	10^9 farads
abhenry	abH	inductance	erg/Bi^2	cm	10^{-9} henry
abohm	abΩ	resistance; impedance; reactance	$erg/Bi^2\ s$	cm/s	10^{-9} ohm
absiemens	abS	conductance; admittance; susceptance	$Bi^2\ s/erg$	s/cm	10^9 siemens
abtesla*	abT	magnetic flux density	$erg/Bi\ cm^2$	$\mathrm{dyn}^{\frac{1}{2}}/cm$	10^{-4} tesla
abvolt	abV	potential (electric); electromotive force	$erg/Bi\ s$	$\mathrm{dyn}^{\frac{1}{2}}\ cm/s$	10^{-8} volt
abweber**	abWb	magnetic flux	erg/Bi	$\mathrm{dyn}^{\frac{1}{2}}\ cm$	10^{-8} weber

* = gauss;   ** = maxwell
Bi = biot; cm = centimetre; dyn = dyne; s = second

ABSOLUTE ...

When international units were employed prior to 1948, the corresponding absolute (ie theoretically defined) units were distinguished from them by the subscript abs. These units are not the same as the absolute electromagnetic units (ab- units). The relationships between absolute and international units are given in the notes accompanying the units involved:

ampère	ohm
coulomb	siemens
farad	tesla
henry	volt
joule	watt
mho	weber

ABSOLUTE DEGREE

See kelvin (2) (note 6)

ABSORPTION UNIT (TOTAL)

Quantity: Equivalent absorption area of a surface to a sound (FPS).

Definition: One (total) absorption unit equals one square foot (ft²) of a surface with a reverberation absorption coefficient of unity, which would absorb sound energy of a given frequency at the same rate as the surface under investigation. 1 (total) absorption unit = 1 ft² = 0·092 903 04 m². 1 m² ≈ 10·763 9 (total) absorption units.
Note: This unit is better called the sabin; it has also been called the open window unit.

ACRE

Quantity: Area (imperial).
Definition: 1 acre = 4 840 yd² = 4 046·856 422 4 m². 1 m² ≈ 2·471 05 × 10⁻⁴ acre.
Note: See Appendix 4 for acre-foot, acre-inch.

AMAGAT DENSITY UNIT

Quantity: Density of a gas (arbitrary).
Definition: One Amagat density unit is the density of a gas in which one mole (mol) occupies a volume of one Amagat volume unit. 1 Amagat density unit = 1 mol/Amagat volume unit ≈ 44·615 8 mol/m³. 1 mol/m³ = 0·022 413 6 Amagat density unit. (These values are on the international scale of atomic masses.)

AMAGAT VOLUME UNIT

Quantity: Volume of a gas (arbitrary).
Definition: One Amagat volume unit is the volume occupied by one mole (mol) of a gas at standard temperature and pressure. The experimentally determined size is 1 Amagat volume unit = (0·022 413 6 ± 0·000 003 0) m³. 1 m³ ≈ 44·615 8 Amagat volume units. (These values are on the international scale of atomic masses.)
Note: The Amagat volume unit has a size numerically equal to the reciprocal of the Amagat density unit.

AMPÈRE

Quantity: Electric current (SI).
Symbol: A; formerly amp.
Definition: The official definition of this basic SI unit, given at the 9CGPM of 1948, is: L'ampère est l'intensité d'un courant constant qui, maintenu dans deux conducteurs parallèles, rectilignes, de longueur infinie, de section circulaire négligeable et placés à une distance de 1 mètre l'un de l'autre dans le vide, produirait entre ces conducteurs une force égale à 2 × 10⁻⁷ unités MKS de force par mètre de longueur. (The ampère is the intensity of a constant current

which, if maintained in two rectilinear parallel conductors of infinite length, of negligible circular cross section, and placed at a distance of 1 metre apart in a vacuum, would produce between these conductors a force equal to 2×10^{-7} newtons per metre of length.)

Note: (1) This unit was formerly called the absolute ampère (A_{abs}) and must be distinguished from the international ampère (A_{int}). The A_{int} was defined at the 4th International Electrical Congress of 1893 as follows: 1 A_{int} is the constant current that, when flowing through a solution of silver nitrate in water under specified conditions, deposits silver at a rate of 0·001 118 grammes per second. This figure was meant to be the numerical value of the electrochemical equivalent of silver in grammes per coulomb (g/C); the experimentally derived value used today is 0·001 118 27 g/C. The definition became law in the UK and the US in 1894, in France (by decree) in 1896, and in Germany in 1898, and remained so until the new definition superceded it in 1948. The CIPM of 1946 formally defined 1 A_{int} = 1·000 34/1·000 49 A $\approx$ 0·999 850 A. 1 A $\approx$ 1·000 15 A_{int}. The international ampère has been called the galvat and the weber.

(2) The standard ampère has been experimentally determined with an Ayrton-Jones current balance to an accuracy (at the National Physical Laboratory) of 4 parts per million.

AMPÈRE, THERMAL

Quantity: Thermal current (SI).
Definition: One thermal ampère corresponds to an entropy flow of one watt per kelvin (W/K). 1 thermal ampère = 1 W/K.
Note: The former definition of the thermal ampère was such that it corresponded to a heat flow rate of one watt.

AMPÈRE-TURN

Quantity: Magnetomotive force (SI).
Symbol: Because the quantity is dimensionally identical with current, the unit symbol A is used. At and AT have also been used.
Definition: One ampère-turn is the magnetomotive force resulting from the passage of a current of one ampère through one turn of a coil.

ÅNGSTRÖM Å

Quantity: Length, especially the wavelength of visible and near-visible electromagnetic radiation (metric).
Definition: 1 Å = 10^{-10} m. 1 m = 10^{10} Å.
Note: (1) The unit symbols A and AU (ångström unit) should not be employed.

(2) The former definition of the ångström is: 1 Å = 1/6 438·469 6 of the wavelength of the red cadmium line in dry air at standard atmospheric pressure, 15 °C and 0·03 % by volume of carbon dioxide. This is now called the international ångström (IÅ), and 1 IÅ = 1·000 000 2 Å.

(3) The unit is sometimes called the tenthmetre.

APOSTILB   asb

Quantity: Luminance (metric).
Definition: One apostilb is the luminance of a uniform diffuser emitting one lumen per metre squared (m²). In terms of the candela (cd), 1 asb = $1/\pi$ cd/m² = $1/\pi$ nit (nt). 1 nt = π asb.
Note: The use of this unit is deprecated: the nit should be used in its place. It has also been called the blondel.

ARCMIN '

Quantity: (Geometrical) plane angle (all).
Definition: $1' = \frac{1}{60}° = \pi/10\ 800$ radian.
Note: This unit is better known as the minute (of arc).

ARE   a

Quantity: Area (metric).
Definition: One are is the area of a square of side length ten metres. 1 a = 100 m². 1 m² = 0·01 a.
Note: This unit is only used in agrarian applications. Generally the hectare is employed, where 1 ha = 100 a = 10 000 m².

A-SIZE

Quantity: A size to which (trimmed) paper and board is manufactured (metric).
Definition: The criteria for forming members of the series are:
(1) If, for any member of the series, x = the length of the shorter sides, and y = the length of the longer sides, the next member is obtained by dividing the previous one into two along the perpendicular bisector of its longer sides. Thus for the next member, length of the longer sides = x, and length of the shorter sides = $\frac{1}{2}y$. Since the shapes of the two members are similar,

$$y : x = x : \tfrac{1}{2}y.$$
Thus, $\qquad\qquad\qquad y : x = \sqrt{2} : 1.$

(2) The basic member of the series has an area of one square metre (m²). Thus, $xy = 1$ m². This gives $x = 2^{-\frac{1}{4}}$ m and $y = 2^{\frac{1}{4}}$ m. These values are conventionally taken as $x = 0·841$ m and $y = 1·189$ m.

Note: (1) When dividing odd numbers by 2 the result is rounded down to the nearest whole number.

(2) Long sizes are produced from the basic series by dividing the sheet into a number of equal parts parallel to the shorter sides.

(3) The manufacturing tolerances allowed are $\pm 1 \cdot 5$ mm for dimensions up to and including 150 mm, ± 2 mm for dimensions greater than 150 mm and up to and including 600 mm, and ± 3 mm for dimensions greater than 600 mm.

(4) Alternative metric series are the B-, C-, RA- and SRA-sizes.

Members of the A-series

designation	size $x \times y$ mm mm		imperial equivalent $x \times y$ in in		notes
4A0	1 682	2 378	66·22	93·62	1
2A0	1 189	1 682	46·81	66·22	1
A0	841	1 189	33·11	46·81	2
A1	594	841	23·39	33·11	
A2	420	594	16·54	23·39	
A3	297	420	11·69	16·54	
A4	210	297	8·27	11·69	
A5	148	210	5·83	8·27	
A6	105	148	4·13	5·83	
A7	74	105	2·91	4·13	
A8	52	74	2·05	2·91	
A9	37	52	1·46	2·05	
A10	26	37	1·02	1·46	
$\frac{1}{3}$A4	99	210	3·90	8·27	3
$\frac{1}{4}$A4	74	210	2·91	8·27	3
$\frac{1}{8}$A7	13	74	0·51	2·91	3

1 Additional, rarely used size
2 Basic member of the A-series
3 Popular long size

ASTRONOMICAL UNIT

Quantity: Length (arbitrary).

Symbol: The unit symbol au is recommended; AU is also used.

Definition: As formally adopted by the International Astronomical Union, 1 au $= 1 \cdot 496\ 00 \times 10^{11}$ m. 1 m $\approx 6 \cdot 684\ 49 \times 10^{-12}$ au.

Note: (1) The definition of the experimentally derived astronomical unit is the mean distance between the earth and the sun, ie the length of the semi-major axis of the earth's orbit. This gives 1 au $= (1 \cdot 495\ 984 \times 10^{11} \pm 2 \cdot 7 \times 10^{5})$ m.

(2) This unit is used for astronomical measurements of distances, normally confined to the solar system.

ATMO-METRE   atmo-m(0)

Quantity: Depth of equivalent atmosphere (arbitrary).

Definition: x atmo-m of gas X is the depth (in metres) that an atmosphere would have if gas X were the only constituent and in the same amount as exists in the actual atmosphere, and reduced to standard temperature and pressure (stp). The number of molecules in unit volume of a gas at stp is given by the ratio of Avogadro's number ($6 \cdot 022\ 52 \times 10^{23}$ molecules/mole, international scale) to one Amagat volume unit: $2 \cdot 686\ 99 \times 10^{25}$ molecules/m³. Thus, 1 atmo-m $\equiv 2 \cdot 686\ 99 \times 10^{25}$ molecules/m². 1 molecule/m² $\approx 3 \cdot 721\ 63 \times 10^{-26}$ atmo-m.

Note: The unit is better called the metre-atmosphere.

ATMOSPHERE   atm

Quantity: Pressure (metric)

Definition: 1 atm = 101 325 pascals (Pa). 1 Pa $\approx 9 \cdot 869\ 23 \times 10^{-6}$ atm.

Note: (1) The name of this unit is more correctly the standard atmosphere. It has also been called (incorrectly) the normal atmosphere.

(2) The above definition was formally adopted by the 10CGPM of 1954 to replace the earlier definition, 1 atm = 760 millimetres of mercury.

(3) The unit must not be confused with the technical atmosphere.

ATMOSPHERE, TECHNICAL   at

Quantity: Pressure (metric).

Definition: One technical atmosphere is the pressure resulting from a force of one kilogramme-force (kgf) acting uniformly over an area of one square centimetre (cm²). 1 at = 1 kgf/cm² = 10^4 kgf/m² = 98 066·5 pascals (Pa). 1 Pa $\approx 1 \cdot 019\ 72 \times 10^{-5}$ at.

Note: The unit must not be confused with the (standard) atmosphere (atm). 1 at $\approx 0 \cdot 967\ 841$ atm; 1 atm $\approx 1 \cdot 033\ 23$ at.

ATOMIC MASS UNIT (UNIFIED)

Quantity: Mass (arbitrary).

Symbol: amu(BS) or u(BS).

(1) Atomic mass unit, international scale   amu(international)

Definition: One amu(international) is equal to one-twelfth of the mass of a neutral carbon-12 atom. The experimentally derived value is 1 amu(international) = $(1 \cdot 660\ 33 \pm 0 \cdot 000\ 05) \times 10^{-27}$ kg. 1 kg $\approx 6 \cdot 022\ 90 \times 10^{26}$ amu(international).

Note: This unit is also called the dalton.

(2) Atomic mass unit, physical scale   amu(physical)

Definition: One amu(physical) is equal to one-sixteenth of the mass of a neutral oxygen-16 atom. The experimentally derived value is 1 amu-(physical) = $(1\cdot659\ 81 \pm 0\cdot000\ 05) \times 10^{-27}$ kg. 1 kg $\approx 6\cdot024\ 79 \times 10^{26}$ amu(physical).

(3) Atomic mass unit, chemical scale   amu(chemical)

Definition: One amu(chemical) is equal to one-sixteenth of the weighted average mass of the three naturally-occurring neutral isotopes of oxygen. The isotopes are $^{16}_{8}O$, $^{17}_{8}O$ and $^{18}_{8}O$, which are found in the ratio of 506 : 0·24 : 1, and thus the experimentally derived value is 1 amu-(chemical) = $(1\cdot660\ 26 \pm 0\cdot000\ 05) \times 10^{-27}$ kg. 1 kg $\approx 6\cdot023\ 15 \times 10^{26}$ amu(chemical).

Note: This unit is also called the chemical mass unit. It was formerly called the atomic weight unit.

Note: (1) 1 amu(international) $\approx$ 1·000 31 amu(physical) $\approx$ 1·000 04 amu(chemical); 1 amu(physical) $\approx$ 0·999 687 amu(international) $\approx$ 0·999 729 amu(chemical); 1 amu(chemical) $\approx$ 0·999 958 amu(international) $\approx$ 1·000 27 amu(physical).

(2) This unit, also called the physical mass unit, must not be confused with the atomic unit of mass.

ATOMIC UNIT OF CHARGE

Definition: One atomic unit of charge is equal to the charge on the electron. The experimentally derived value is 1 atomic unit of charge = $(1\cdot602\ 10 \pm 0\cdot000\ 07) \times 10^{-19}$ coulomb (C). 1 C $\approx 6\cdot241\ 81 \times 10^{18}$ atomic units of charge.

Note: This is a fundamental unit of the Hartree system of units (see Appendix 1 section 10).

ATOMIC UNIT OF ENERGY

(1) *Definition:* The preferred definition is 1 atomic unit of energy = e^2/a_0, where e is the atomic unit of charge and a_0 is the atomic unit of length. Hence 1 atomic unit of energy $\approx 4\cdot850\ 5 \times 10^{-18}$ joule (J). 1 J $\approx 2\cdot061\ 6 \times 10^{17}$ atomic units of energy.

Note: (1) This unit has also been called the hartree.

(2) *Definition:* An alternative definition is 1 atomic unit of energy = $e^2/2a_0 \approx$ 2·425 2 $\times 10^{-18}$ J. 1 J $\approx 4\cdot123\ 3 \times 10^{17}$ atomic units of energy.

Note: (2) This unit is also called the rydberg.

Note: (3) This is a derived unit of the Hartee system of units (see Appendix 1 section 10).

ATOMIC UNIT OF LENGTH

Definition: One atomic unit of length is equal to the radius of the first Bohr orbit (for infinite mass). The experimentally derived value is 1 atomic unit of length $= (5\cdot291\ 67 \pm 0\cdot000\ 07) \times 10^{-11}$ metre (m). $1\ m \approx 1\cdot889\ 76 \times 10^{10}$ atomic units of length.

Note: This is a derived unit of the Hartree system of units (see Appendix 1 section 10).

ATOMIC UNIT OF MASS

Definition: One atomic unit of mass is equal to the rest mass of the electron. The experimentally derived value is 1 atomic unit of mass $= (9\cdot108\ 4 \pm 0\cdot000\ 3) \times 10^{-31}$ kg. $1\ kg \approx 1\cdot097\ 9 \times 10^{30}$ atomic units of mass.

Note: This is a fundamental unit of the Hartree system of units (see Appendix 1 section 10). It must not be confused with the atomic mass unit.

ATOMIC UNIT OF TIME

Definition: One atomic unit of time is equal to the period of the first Bohr orbit. The experimentally derived value is 1 atomic unit of time $= (2\cdot418\ 84 \pm 0\cdot000\ 003) \times 10^{-17}$ second (s). $1\ s \approx 4\cdot134\ 21 \times 10^{16}$ atomic units of time.

Note: This is a derived unit of the Hartree system of units (see Appendix 1 section 10).

ATOMIC WEIGHT UNIT   awu(BS)

Quantity: Mass (arbitrary).

Definition: One atomic weight unit is equal to one-sixteenth of the weighted average mass of the three naturally occurring neutral isotopes of oxygen. The isotopes are $^{16}_{8}O$, $^{17}_{8}O$ and $^{18}_{8}O$, which are found in the ratio of $506 : 0\cdot24 : 1$, and thus the experimentally derived value is 1 atomic weight unit $= (1\cdot660\ 26 \pm 0\cdot000\ 05) \times 10^{-27}$ kg. $1\ kg \approx 6\cdot023\ 15 \times 10^{26}$ atomic weight units.

Note: (1) The unit is better called the atomic mass unit (chemical scale). It is also called the chemical mass unit.

(2) The atomic mass unit (international) is preferred to the atomic weight unit.

ATTO-...

See Appendix 3, section 2. Units are not entered under the prefix atto-.

BALMER

Quantity: Reciprocal length, especially wave number (CGS).

Definition: One balmer is the reciprocal length of a distance which has a

length of one centimetre (cm). 1 balmer = 1/cm = 100/m. 1 m = 100/balmer.
Note: This unit is better called the kayser. It has also been called the rydberg.

BAR

Quantity: Pressure (metric).
Symbol: In meteorology the unit symbol b is used in submultiples; otherwise no unit symbol is used.
Definition: 1 bar = 10^5 pascals (Pa). 1 Pa = 10^{-5} bar.
Note: (1) The basic unit is rarely employed, but the sub-unit the millibar (mb = 0·001 bar = 100 Pa).
(2) The unit was at one time also used with the definition 1 bar = 1 dyne per centimetre squared (CGS unit), but this should be called the barye.

BARAD

Quantity: Pressure (CGS).
Definition: One barad is the pressure resulting from a force of one dyne (dyn) acting uniformly over an area of one square centimetre (cm^2). 1 barad = 1 dyn/cm^2 = 0·1 pascal (Pa). 1 Pa = 10 barads.
Note: The barad is a former name for the barye (dyn/cm^2 or microbar), which should always be employed. It has also been called the rum.

BARN   b(BS)

Quantity: Area, especially the cross-sectional area of an atomic nucleus (metric).
Definition: 1 b = 10^{-28} m^2. 1 m^2 = 10^{28} b.

BARREL, DRY   bbl(O)

Quantity: Capacity, ie volume (imperial).
Definition: 1 bbl = 7 056 in^3 = 0·115 627 123 584 m^3. 1 m^3 ≈ 8·648 49 bbl.
Note: This is a US unit used only for the measurement of solid substances; there is no corresponding UK unit. See Appendix 4 for barrel, barrel bulk.

BARYE

Quantity: Pressure (CGS).
Definition: One barye is the pressure resulting from a force of one dyne (dyn) acting uniformly over an area of one square centimetre (cm^2). 1 barye = 1 dyn/cm^2 = 0·1 pascal (Pa). 1 Pa = 10 baryes.
Note: (1) The name barye is uncommon; the unit is more often called the dyn/cm^2 or the microbar.
(2) The unit was formerly called the barad. It has also been called the rum.
(3) The unit was at one time also used with the definition 1 barye = 10^6 dyn/cm^2, but this should be called the bar.

BASE BOX

Quantity: Area (imperial).
Definition: 1 base box = 31 360 square inches = 20·232 217 6 m².
1 m² $\approx$ 0·049 426 1 base box.
Note: This unit is used in metal plating.

BAUD (1)

Quantity: Information for a digital computer (none).
Definition: 1 baud = 1 binary digit.
Note: (1) This unit is better called the bit.
(2) The binary digits are generally represented by the digits 0 and 1.

BAUD (2)

Quantity: Telegraph signalling rate (all).
Definition: 1 baud = 1 pulse per second.

B-DOSE

Quantity: Radiation dose (arbitrary).
Definition: One B-dose is the dose of radiation required to change the colour
of a barium platinocyanide pastille from a specified apple-green colour (tint
'A') to a specified red-brown (tint 'B').
Note: This obsolete unit is better called the pastille dose. It is equal to 500
röntgens approximately.

BEL

Quantity: Intensity level; in the special case explained in note (3) the unit
measures amplitude level and logarithmic decrement (all).
Symbol: B. The unit symbol b is generally used in telecommunications
technology.
Definition: The intensity level N in bels relates two powers P_1, P_2 (expressed
in the same units) by the formula $N = \log (P_1/P_2)$. Thus 1 bel is the intensity
level that corresponds to a power ratio of 10.
Note: (1) The basic unit is never employed, but the sub-unit the decibel
(dB or db): 1 dB = 0·1 B.
(2) The names decilit, decilog, decilu, decomlog, logit and transmission unit
have been used at various times for the decibel.
(3) Under those conditions in which the square of the amplitude of a
vibration is proportional to the associated power there is a connexion between
the decibel and the neper (Np): 1 dB = $\frac{1}{20}$ ln 10 Np $\approx$ 0·115 129 Np.

BENZ

Quantity: Velocity (SI).
Definition: 1 benz = 1 metre per second.
Note: This unit has been proposed by Germany, but has not received general acceptance.

BES

Quantity: Mass (CGS).
Definition: 1 bes = 1 gramme.
Note: This name (like the stathm and the brieze) was proposed as an alternative for the gramme, but never employed. In Italy it has been proposed as an alternative for the kilogramme.

BEVA-...

See Appendix 3, section 2. Units are not entered under the prefix beva-.

BICRON

Quantity: Length (metric).
Definition: 1 bicron = 10^{-12} m. 1 m = 10^{12} bicrons.
Note: This little-used unit is also called the stigma.

BIOT   Bi

Quantity: Current (CGSm).
Definition: One biot is the constant current which, if maintained in two rectilinear parallel conductors of infinite length, of negligible cross section, and placed at a distance of one centimetre apart in a vacuum, would produce between these conductors a force equal to two dynes per centimetre of length. 1 Bi = 10 ampères (A). 1 A = 0·1 Bi. The CGS-emu mechanical equivalent is 1 Bi = 1 dyne$^{\frac{1}{2}}$. The CGS-emu equivalent is 1 Bi = 1 abampère (abA).

BIT

Quantity: Information for a digital computer (none).
Definition: 1 bit = 1 binary digit.
Note: (1) The unit has also been called the baud.
(2) The binary digits are generally represented by the digits 0 and 1.

BLINK

Quantity: Time (arbitrary).
Definition: 1 blink = 10^{-5} day = 0·864 s. 1 s $\approx$ 1·157 41 blink.

BLONDEL

Quantity: Luminance (metric).
Definition: One blondel is the luminance of a uniform diffuser emitting one lumen per metre squared (m²). In terms of the candela (cd):

1 blondel $= \dfrac{1}{\pi}$ cd/m² $= \dfrac{1}{\pi}$ nit (nt). 1 nt $= \pi$ blondel.

Note: This unit is better called the apostilb. The nit should always be used in its place.

BOARD OF TRADE UNIT

Quantity: Energy, particularly electrical energy (metric).
Definition: One Board of Trade unit is the energy expended when a power of one kilowatt (kW) is available for one hour (h). Since 1 kW = 1 000 W and 1 h = 3 600 second (s), 1 Board of Trade unit = 1 kW h = $3 \cdot 6 \times 10^6$ joules (J). 1 J $\approx 2 \cdot 777\ 78 \times 10^{-7}$ Board of Trade unit.
Note: This unit is better called the kilowatt-hour.

BOLE

Quantity: Momentum (CGS).
Definition: One bole is the momentum possessed by a mass of one gramme (g) travelling with a velocity of one centimetre per second (cm/s).
1 bole = 1 g cm/s = 10^{-5} kg m/s. 1 kg m/s = 10^5 boles.
Note: This unit has almost never been employed.

BOUGIE DÉCIMALE   bd(O)

Quantity: Luminous intensity (all).
Definition: One bougie décimale is of such a size that the luminance of one square centimetre of surface of molten platinum at its temperature of solidification is 20 bd.
Note: (1) This definition was formulated at the International Electrical Congress of 1889, and was made legal in France by the decree of 1919. The bougie décimale (decimal candle) was subsequently replaced by the candela.
(2) 1 bd = 0·96 international standard candles (measured comparison).

BREWSTER   B(O)

Quantity: Stress optical coefficient of a material, ie reciprocal stress (metric).
Definition: The stress optical coefficient C in brewsters is related to the tensile stress σ in bars which produces a relative retardation s in ångströms (Å) when light passes through a thickness t in millimetres in a direction perpendicular to the stress by the formula $C = s/\sigma t$. Since 1 bar = 10^5 pascals (Pa),

$1 \text{ Å} = 10^{-10}$ m and $1 \text{ mm} = 10^{-3}$ m, $1 \text{ B} = (10^{-10}/10^5 \; 10^{-3})/\text{Pa} = 10^{-12}/\text{Pa}$. $1 \text{ Pa} = 10^{-12}/\text{B}$.

BRIEZE

Quantity: Mass (CGS).
Definition: 1 brieze = 1 gramme.
Note: This name (like the bes and the stathm) was proposed as an alternative for the gramme, but never employed.

BRIG

Quantity: Logarithmic value (all).
Definition: For a value 10^x, the size of the value is given as x brig. x brig corresponds to antilog x.
Note: (1) The brig may be related to the decibel (dB), the neper (Np) and the octave: 1 brig $= 10$ dB, 1 dB $= 0.1$ brig; 1 brig $= \ln 10$ Np $\approx 2.302\,59$ Np, 1 Np $\approx 0.434\,294$ brig; 1 brig $= (1/\log 2)$ octave $\approx 3.321\,93$ octave, 1 octave $\approx 0.301\,030$ brig.
(2) The unit is also called the dex.

BRIL

Quantity: Luminance level, ie subjective luminance (none).
Definition: The luminance level I in brils is related to the luminance L in millilamberts by the formula $L = (\frac{1}{2})^{(100-I)}$. Thus 100 brils is the luminance level that corresponds to a luminance of 1 millilambert.
Note: A procedure whereby the reduction of a stimulus to one-half its value results in the measured quantity being reduced by one step is known as a halving method.

BRITISH THERMAL UNIT Btu

Quantity: Heat energy.
(1) International table British thermal unit Btu$_{IT}$ (imperial).
   Definition: By formal adoption, 1 Btu$_{IT}$ per pound (lb) $= 2\,326$ joules per kilogramme (J/kg). Since 1 lb $= 0.453\,592\,37$ kg, 1 Btu$_{IT} = 2\,326 \times 0.453\,592\,37$ J $= 1\,055.055\,852\,62$ J. $1 \text{ J} \approx 9.478\,17 \times 10^{-4}$ Btu$_{IT}$.
   Note: (1) This unit as defined was adopted by the 5th International Conference on the Properties of Steam (London, July 1956). The definition was chosen in such a way that the values of a specific heat capacity in international table kilocalories per kilogramme kelvin (kcal$_{IT}$/kg K) and international table British thermal units per pound degree Fahrenheit (Btu$_{IT}$/lb degF) are equal in size. 1 kcal$_{IT} = 4\,186.8$ J and 1 K $= \frac{9}{5}$ degF; $4\,186.8 \times \frac{5}{9} = 2\,326$.

(2) Sixty degrees Fahrenheit British thermal unit $Btu_{60/61}$ (arbitrary).

Definition: 1 $Btu_{60/61}$ is the quantity of heat required to raise the temperature of one pound of air-free water from 60 °F to 61 °F at a constant pressure of one (standard) atmosphere. The experimentally derived value is 1 $Btu_{60/61} = 1\ 054 \cdot 5$ J. 1 J $\approx 9 \cdot 483\ 2 \times 10^{-4}\ Btu_{60/61}$.

Note: (2) This unit was in use prior to the Btu_{IT}.

(3) Mean British thermal unit Btu_{mean} (arbitrary).

Definition: 1 Btu_{mean} is $\frac{1}{180}$ of the quantity of heat required to raise the temperature of one pound of air-free water from 32 °F to 212 °F at a constant pressure of one (standard) atmosphere. The experimentally derived value is 1 $Btu_{mean} = 1\ 055 \cdot 8$ J. 1 J $\approx 9 \cdot 471\ 5 \times 10^{-4}\ Btu_{mean}$.

Note: (3) This unit was sometimes used as an alternative to the $Btu_{60/61}$ prior to the Btu_{IT}.

Note: (4) 1 $Btu_{IT} \approx 1 \cdot 000\ 5\ Btu_{60/61} \approx 0 \cdot 999\ 30\ Btu_{mean}$;
1 $Btu_{60/61} \approx 0 \cdot 999\ 47\ Btu_{IT} \approx 0 \cdot 998\ 77\ Btu_{mean}$;
1 $Btu_{mean} \approx 1 \cdot 000\ 7\ Btu_{IT} \approx 1 \cdot 001\ 2\ Btu_{60/61}$.

(5) Former unit symbols used were BThU and BTU.

(6) Former units include the thirty-nine degrees Fahrenheit British thermal unit ($Btu_{39/40}$), being the quantity of heat required to raise the temperature of one pound of air-free water from 39 °F to 40 °F at a constant pressure of one (standard) atmosphere. The experimentally derived value is 1 $Btu_{39/40} = 1\ 059 \cdot 52$ J.

B-SIZE

Quantity: A size to which (trimmed) paper and board is manufactured (metric).

Definition: The criteria for forming members of the series are:

(1) If, for any member of the series, $x = $ the length of the shorter sides, and $y = $ the length of the longer sides, the next member is obtained by dividing the previous one into two along the perpendicular bisector of its longer sides. Thus for the next member, length of the longer sides $= x$, and length of the shorter sides $= \frac{1}{2}y$. Since the shapes of the two members are similar,

$$y : x = x : \tfrac{1}{2}y.$$
Thus, $$y : x = \sqrt{2} : 1.$$

(2) The basic member of the series has an area of $\sqrt{2}$ square metres (m²) Thus, $xy = \sqrt{2}$ m². This gives $x = 1$ m and $y = \sqrt{2}$ m. The value of the latter is conventionally taken as $y = 1 \cdot 414$ m.

Note: (1) When dividing odd numbers by 2 the result is rounded down to the nearest whole number.

(2) The manufacturing tolerances allowed are $\pm 1 \cdot 5$ mm for dimensions up to and including 150 mm, ± 2 mm for dimensions greater than 150 mm and

up to and including 600 mm, and ± 3 mm for dimensions greater than 600 mm.

(3) Alternative metric series are the A-, C-, RA- and SRA-sizes.

Members of the B-series

designation	size x × y mm mm		imperial equivalent x × y in in	
B0	1 000	1 414	39·37	55·67
B1	707	1 000	27·83	39·37
B2	500	707	19·68	27·83
B3	353	500	13·90	19·68
B4	250	353	9·84	13·90
B5	176	250	6·93	9·84
B6	125	176	4·92	6·93
B7	88	125	3·46	4·92
B8	62	88	2·44	3·46
B9	44	62	1·73	2·44
B10	31	44	1·22	1·73

BUSHEL   bu (BS)

Quantity: Capacity, ie volume.
Definition: (1) UK unit (arbitrary)
    1 UKbu = 8 UK gallons = 64 UK pints $\approx$ 0·036 368 7 m³.
    1 m³ $\approx$ 27·496 2 UKbu.
(2) US unit (imperial)
    1 USbu = 2 150·42 in³ = 64 US dry pints = 0·035 239 070 166 88 m³.
    1 m³ $\approx$ 28·377 6 USbu.
Note: (1) The UK bushel is used for the measurement of solid and liquid substances, although usually the former; the US bushel is used only for the measurement of solid substances. 1 UKbu $\approx$ 1·032 06 USbu; 1 USbu $\approx$ 0·968 939 UKbu.
(2) The bushel has also been employed with local definitions different from those given above.
(3) See Appendix 4 for another entry 'bushel' and international corn bushel.

BYTE

Quantity: Information for a digital computer (none).
Definition: 1 byte = 8 binary digits = 8 bits. 1 bit = 0·125 byte.
Note: The binary digits are generally represented by the digits 0 and 1.

CABLE (LENGTH)

Quantity: Length, particularly for marine use (imperial).
Definition: The unit has no definitive size, the following being the more common definitions.
(1) 1 cable length = 120 fathoms = 720 ft = 219·456 m.
   1 m $\approx$ 4·556 72 $\times$ 10^{-3} cable length.
(2) 1 cable length = $\frac{1}{10}$ UK nautical mile = 608 ft = 185·318 m.
   1 m $\approx$ 5·396 12 $\times$ 10^{-3} cable length.
(3) 1 cable length = 100 fathoms = 600 ft = 182·88 m.
   1 m $\approx$ 5·468 07 $\times$ 10^{-3} cable length.

CALORIE   cal

Quantity: Heat energy.
(1) International table calorie cal$_{IT}$ (metric-derived).
   Definition: By formal adoption, 1 cal$_{IT}$ = 4·186 8 joules (J).
   1 J $\approx$ 0·238 846 cal$_{IT}$.
   Note: (1) This unit was adopted with the above definition by the Fifth International Conference on the Properties of Steam (London, July 1956). It was generally regarded as the best definition of the calorie.
(2) Fifteen-degrees calorie (15 °C calorie) cal$_{15}$ (arbitrary).
   Definition: One fifteen-degrees calorie is the quantity of heat required to raise the temperature of one gramme of air-free water from 14·5 °C to 15·5 °C at a constant pressure of one (standard) atmosphere. The experimentally-derived value is 1 cal$_{15}$ = 4·185 5 $\pm$ 0·000 5 J. 1 J $\approx$ 0·238 92 cal$_{15}$.
   Note: (2) The fifteen-degrees calorie, also called the gramme-calorie and the small calorie, was adopted by the International Union of Pure and Applied Physics in 1934. The formal definition was proposed by the Comité Consultatif de Thermométrie et Calorimétrie, and adopted by the CIPM in 1950 as being the most accurately-determined experimental value.
   (3) The large calorie is defined by 1 large calorie = 1 000 cal$_{15}$. (The unit symbol Cal, spelled with a capital letter, is deprecated and should be avoided.) The unit is better called the kilocalorie; it has also been called the kilogramme-calorie.
   (4) The fifteen-degrees calorie was in use prior to the international table calorie.
(3) Thermochemical calorie cal$_{tc}$ (metric-derived).
   Definition: By formal adoption, 1 cal$_{tc}$ = 4·184 J. 1 J $\approx$ 0·239 006 cal$_{tc}$.
   Note: (5) This calorie has been used in thermochemistry, in preference to the others defined above.

Note: (6) 1 $cal_{IT} \approx 1 \cdot 000\ 31\ cal_{15} \approx 1 \cdot 000\ 67\ cal_{tc}$;

      1 $cal_{15} \approx 0 \cdot 999\ 690\ cal_{IT} \approx 1 \cdot 000\ 36\ cal_{tc}$;

      1 $cal_{tc} \approx 0 \cdot 999\ 331\ cal_{IT} \approx 0 \cdot 999\ 642\ cal_{15}$.

(7) The 9CGPM of 1948 adopted the joule as the unit of measurement of heat energy, and recommended that the calorie be avoided as far as possible. The unit is now obsolete.

(8) Former units include:

(a) The four-degrees calorie (cal_4). The quantity of heat required to raise the temperature of one gramme of air-free water from 3·5 °C to 4·5 °C at a constant pressure of one (standard) atmosphere. The experimentally-derived value is 1 $cal_4 = 4 \cdot 204\ 5$ J.

(b) The mean calorie (cal_{mean}). One-hundreth of the quantity of heat required to raise the temperature of one gramme of air-free water from 0 °C to 100 °C at a constant pressure of one (standard) atmosphere. The experimentally-derived value is 1 $cal_{mean} = 4 \cdot 189\ 7$ J.

CANDELA   cd

Quantity: Luminous intensity (all).

Definition: The official definition of this basic SI unit, given at the 9CGPM of 1948 (and called at that time the new candle) is: La grandeur de la bougie nouvelle est telle que la brilliance du radiateur intégral, à la température de solidification du platine, soit de 60 bougies nouvelles par centimètre carré. (The magnitude of the candela is such that the luminance of a full radiator at the temperature of solidification of platinum is 60 candelas per square centimetre.)

Note: On the international temperature scale of 1968 the temperature of solidification of platinum is 2 045 K.

CANDLE, DECIMAL

See bougie décimale (note 1)

CANDLE-FOOT

See foot-candle (note)

CANDLE, INTERNATIONAL STANDARD

Quantity: Luminous intensity (all).

Definition: One international standard candle is the average luminous intensity of the candle standards of the UK, the US and France. The average measured value is $58 \cdot 9 \pm 0 \cdot 2$ candles = 60 candelas (cd), the actual measured value for the UK being 59·0 candles = 60 cd.

Note: (1) The international standard candle originally had the following definitions:

- (a) (Before 1800) The mean intensity of the British standard candle. The British standard candle was made of spermaceti wax, weighed one-sixth of a pound, and burned at a rate of 120 grains per hour.
- (b) (About 1800) Approximately one-tenth of the intensity of the Carcel lamp. See the entry carcel.
- (c) (1877) Approximately, the intensity of the Vernon Harcourt pentane vapour lamp.

(2) During the first half of the nineteenth century other pentane-burning lamps were employed as standard, and in particular the Dibdin lamp (ten candles approximately) and the Simmance lamp (two candles approximately). (3) All the above lamps were subject to variation because of changes in atmospheric humidity and pressure, and their intensities were defined under certain conditions.

CANDLE, NEW   cd

Quantity: Luminous intensity (all).
Definition: The official definition of this basic SI unit, given at the 9CGPM is: La grandeur de la bougie nouvelle est telle que la brilliance du radiateur intégral, à la température de solidification du platine, soit de 60 bougies nouvelles par centimètre carré. (The magnitude of the new candle is such that the luminance of a full radiator at the temperature of solidification of platinum is 60 new candles per square centimetre.)
Note: (1) On the international temperature scale of 1968 the temperature of solidification of platinum is 2 045 K.
(2) The name of the unit is now the candela.

CARAT

Quantity: Mass (imperial).
Definition: 1 carat = 4 grains = $2 \cdot 591\ 956\ 4 \times 10^{-4}$ kg.
1 kg $\approx$ 3 858·09 carats.
Note: This is an obsolete unit of troy measure, now replaced by the metric carat.

CARAT, METRIC

Quantity: Mass (metric).
Definition: 1 metric carat = 200 mg = 2×10^{-4} kg.
1 kg = 5 000 metric carats.
Note: This unit was adopted by the 4CGPM of 1907 for commerical transactions in diamonds, pearls and other precious stones. It is the only carat legally acceptable in the UK.

CARCEL

Quantity: Luminous intensity (all).
Definition: 1 carcel = 9·61 international standard candles (measured comparison).
Note: This is an obsolete unit. The Carcel lamp itself burned colza oil, and the intensities of different lamps varied between 9·4 and 10 international standard candles.

CASCADE UNIT

Quantity: Length in cosmic-ray studies (arbitrary).
Definition: 1 cascade unit = ln 2 shower units ≈ 0·693 147 shower units (qv).
Note: (1) This unit is also called the radiation length or the radiation unit.
(2) The cascade unit is an individual unit of length, ie its size depends on the conditions of a given set of circumstances.

CÉ

Quantity: Time (arbitrary).
Definition: 1 cé = 0·01 day = 864 seconds (s). 1 s 0·001 ≈ 157 41 cé.
Note: A metric unit suggested for the measurement of time, but almost never employed. It has also been called the degree.

CELO

Quantity: Acceleration (FPS).
Definition: One celo is the acceleration of a body, the velocity of which changes by one foot per second (ft/s) in one second.
1 celo = 1 ft/s² = 0·304 8 m/s². 1 m/s² ≈ 3·280 84 celos.

CELSIUS DEGREE

Quantity: Temperature interval or difference (SI).
Symbol: deg; the unit symbol was formerly degC.
Definition: 1 deg = 1 kelvin.
Note: (1) The original definition was: 1 deg is one-hundreth of the interval between the freezing and boiling points of pure air-free water, both under a pressure of one (standard) atmosphere.
(2) The unit was formerly the Centigrade degree. The use of the term centigrade should be restricted to the unit 0·01 grade.
(3) See also the entry degree Celsius.

CENT   (1)

Quantity: (Pitch) interval (all).
Definition: The pitch interval I_c in cents between two frequencies f_1, f_2 (expressed in the same units, and with f_2 greater than f_1) is given by the formula $I_c = \dfrac{1\ 200}{\log 2} \log (f_2/f_1) \approx 3\ 986\cdot31 \log (f_2/f_1)$.
Note: I_c is related to the corresponding pitch interval I_o in octaves by $I_c = 1\ 200\ I_o$; $I_o \approx 0\cdot083\ 333\ 3\ I_c$.

CENT   (2)

Quantity: Reactivity (a measure of the departure of a nuclear reactor from its critical condition) (all).
Definition: 1 cent = 0·01 dollar (qv).
Note: This unit is rarely used.

CENTAL   ctl (BS)

Quantity: Mass (imperial).
Definition: 1 ctl = 100 lb = 45·359 237 kg. 1 kg $\approx$ 0·022 046 2 ctl.
Note: This is the UK name for the unit; in the US it is called the short hundredweight, or just the hundredweight. It is also called the centner and the quintal.

CENTI-. . .

See Appendix 3, section 2. The only units referred to in Part 1 are the following:
    centigrade: see Celsius degree (note 2)
    centimetre*
    centimetre-candle*
    centimetre of mercury (conventional)*
* Entered in its alphabetical position

CENTIGRADE DEGREE

See Celsius degree (note 2)

CENTIGRADE HEAT UNIT

Quantity: Heat energy (arbitrary).
Symbol: CHU; more correctly the unit symbol should be CHU$_{mean}$.
Definition: One centigrade heat unit is one-hundredth of the quantity of heat required to raise the temperature of one pound of air-free water from 0 °C to 100 °C at a constant pressure of one (standard) atmosphere. Since 1 degC =

$\frac{9}{5}$ degF, 1 CHU = $\frac{9}{5}$ mean British thermal unit (Btu$_{mean}$). The experimentally derived size of the Btu$_{mean}$ is 1 Btu$_{mean}$ = 1 055·8 joules (J), and thus 1 CHU = $\frac{9}{5}$ × 1 055·8 J = 1 900·44 J. 1 J ≈ 5·261 94 × 10^{-4} CHU.

Note: The full name for this unit is the mean pound Centigrade (or, more correctly, Celsius) heat unit.

CENTIHG

Quantity: Pressure (metric).

Definition: One centihg is the pressure that would support a column of mercury of length one centimetre and density 13 595·1 kilogrammes per cubic metre under the standard acceleration of free fall (g_n). Since g_n = 9·806 65 m/s^2, 1 centihg = 1 cmHg = 13 595·1 × 9·806 65 × 10^{-2} pascal (Pa) = 1 333·223 874 15 Pa. 1 Pa ≈ 0·000 750 062 centihg.

Note: The unit is better called the (conventional) centimetre of mercury. It is equal in size to the decatorr, to one part in seven million. The name is pronounced with the *h* silent.

CENTIMETRE   (1)   cm

Quantity: Capacitance (CGS-esu).

Definition: One centimetre is the capacitance of a condenser having a charge of one statcoulomb (statC), across the plates of which the potential difference is one statvolt (statV). 1 cm = 1 statC/statV = 10^5/$(c)^2$ farad (F), where (c) is the numerical value of the velocity of electromagnetic radiation in vacuo expressed in metres per second. Since the experimentally derived value of (c) is 2·997 925 × 10^8, 1 cm ≈ 1·112 65 × 10^{-12} F. 1 F ≈ 8·987 55 × 10^{11} cm.

Note: (1) An alternative names for the unit is the statfarad.

(2) In the US the name of this unit is spelled centimeter. The spelling in official translations of ISO Recommendations is always centimetre.

CENTIMETRE   (2)   cm

Quantity: Inductance (CGS-emu)

Definition: One centimetre is the inductance of a coil in a closed circuit which gives rise to a magnetic flux of one maxwell (Mx) per abampère (abA). 1 cm = 1 Mx/abA = 10^{-9} henry (H). 1 H = 10^9 cm.

Note: (1) An alternative name for the unit is the abhenry.

(2) In the US the name of this unit is spelled centimeter. The spelling in official translations of ISO Recommendations is always centimetre.

CENTIMETRE   (3)   cm

Quantity: Length (CGS, Mie).
Definition: 1 cm = 0·01 m. 1 m = 100 cm.
Note: In the US the name of this unit is spelled centimeter. The spelling in official translations of ISO Recommendations is always centimetre.

CENTIMETRE-CANDLE   cm-c (O)

Quantity: (Intensity of) illumination (CGS).
Definition: One centimetre-candle is the illumination of one lumen (lm) uniformly over an area of one square centimetre (cm^2).
1 cm-c = 1 lm/cm^2 = 10^4 lux (lx). 1 lx = 10^{-4} cm-c.
Note: The name centimetre-candle is deprecated: the phot should be used in its place.

CENTIMETRE OF MERCURY (CONVENTIONAL)   cmHg

Quantity: Pressure (metric).
Definition: One (conventional) centimetre of mercury is the pressure that would support a column of mercury of length one centimetre and density 13 595·1 kilogrammes per cubic metre under the standard acceleration of free fall (g_n). Since g_n = 9·806 65 m/s^2, 1 cmHg = 13 595·1 × 9·806 65 × 10^{-2} pascal (Pa) = 1 333·223 874 15 Pa. 1 Pa ≈ 0·000 750 062 cmHg.
Note: This unit has also been called the centihg. It is equal in size to the decatorr, to one part in seven million.

CENTNER

Quantity: Mass (imperial).
Definition: 1 centner = 100 lb = 45·359 237 kg. 1 kg ≈ 0·022 046 2 centner.
Note: In the US this unit is called the short hundredweight, or just hundredweight. It has also been called the quintal and the cental.

CENTNER, METRIC

Quantity: Mass (metric).
Definition: There are two alternative definitions.
(1) 1 metric centner =   50 kg. 1 kg = 0·02 metric centner.
(2) 1 metric centner = 100 kg. 1 kg = 0·01 metric centner.
Note: The unit corresponding to the second of these definitions is also called the quintal.

CENTRAD

Quantity: (Geometrical) plane angle, particularly angular deviation of light through a prism (all).
Definition: 1 centrad = 0·01 radian (rad). 1 rad = 100 centrads.

CHAD

Quantity: Neutron flux (metric).
Definition: There are two alternative definitions:
(1) One chad is a neutron flux of one neutron per square centimetre (cm^2) per second (s). 1 chad = 1 neutron/cm^2 s = 10^4 neutrons/m^2 s. 1 neutron/m^2 s = 10^{-4} chad.
(2) One chad is a neutron flux of 10^{12} neutrons per square centimetre per second. 1 chad = 10^{12} neutrons/cm^2 s = 10^{16} neutrons/m^2 s. 1 neutron/m^2 s = 10^{-16} chad.
Note: This unit has rarely been employed.

CHAIN (1)

Quantity: Area (imperial).
Definition: 1 chain = 484 square yards = 404·685 642 24 m^2. 1 m^2 ≈ 0·002 471 05 chain.
Note: This unit is more strictly called the square chain.

CHAIN (2) ch (O)

Quantity: Length (imperial).
Definition: 1 ch = 22 yards = 20·116 8 m. 1 m ≈ 0·049 709 7 ch.
Note: This is the legally defined chain, commonly called Gunter's chain in the US. It must be distinguished from the engineer's chain and the nautical chain.

CHAIN, ENGINEER'S

Quantity: Length (imperial).
Definition: 1 engineer's chain = 100 feet = 30·48 m. 1 m ≈ 0·032 808 4 engineer's chain.

CHAIN, NAUTICAL

Quantity: Length (imperial).
Definition: 1 nautical chain = 15 feet = 4·572 m. 1 m ≈ 0·218 723 nautical chain.

CHALDRON

Quantity: Capacity, ie volume.
Definition: (1) UK unit (arbitrary)
   1 UK chaldron = 36 UK bushels = 288 UK gallons $\approx$ 1·309 27 m³.
   1 m³ $\approx$ 0·763 783 UK chaldron.
(2) US unit (imperial)
   1 US chaldron = 36 US bushels = 1·268 606 526 007 68 m³.
   1 m³ $\approx$ 0·788 266 US chaldron.
Note: (1) The UK chaldron is used for the measurement of solid and liquid substances; the US chaldron, before it became obsolete, was used only for the measurement of solid substances. 1 UK chaldron $\approx$ 1·032 06 US chaldron; 1 US chaldron $\approx$ 0·968 939 UK chaldron.
(2) See also the entry in Appendix 4.

CHEMICAL MASS UNIT

Quantity: Mass (arbitrary).
Definition: One chemical mass unit is equal to one-sixteenth of the weighted average mass of the three naturally occurring neutral isotopes of oxygen. The isotopes are $^{16}_{8}O$, $^{17}_{8}O$ and $^{18}_{8}O$, which are found in the ratio of 506:0·24:1 and thus the experimentally derived value is 1 chemical mass unit = $(1\cdot660\ 26 \pm 0\cdot000\ 05) \times 10^{-27}$ kg. 1 kg $\approx$ 6·023 15 $\times$ 10²⁶ chemical mass units.
Note: (1) The unit is better called the atomic mass unit (chemical scale). It was formerly called the atomic weight unit.
(2) The atomic mass unit (international) is preferred to the chemical mass unit.

CIRCLE

Quantity: (Geometrical) plane angle (all).
Definition: 1 circle = 2π radians = 360°.
Note: This unit is also called the turn.

CLAUSIUS

Quantity: Entropy (arbitrary).
Definition: One clausius is the entropy associated with a temperature of one kelvin (K) in which there is an increase in heat of one thousand calories (cal$_{IT}$).
1 clausius = 1 000 cal$_{IT}$/K = 4 186·8 joules per kelvin (J/K).
1 J/K $\approx$ 2·388 46 $\times$ 10⁻⁴ clausius.

CLO

Quantity: Thermal insulation of clothing (arbitrary).
Definition: The unit is formally defined by 1 clo = 0·875 foot hour degree-Fahrenheit per international table British thermal unit. 1 clo ≈ 0·505 566 metre kelvin per watt (m K/W). 1 m K/W ≈ 1·977 98 clo.
Note: This US unit was originally defined as the insulation required to maintain a stationary person at a comfortable temperature under indoor conditions.

CLUSEC

Quantity: Leak rate, ie power (metric-derived).
Definition: One clusec is a leak rate of one centilitre (cl) per second (s), at a pressure of one millitorr. 1 clusec = 0·01 × 0·001 l torr/s. Using the new litre, 0·001 m³, and with 1 torr ≈ 133·322 pascals, 1 clusec ≈ 0·000 01 × 0·001 × 133·322 watt (W) ≈ 1·333 22 × 10^{-6} W. 1 W ≈ 7·500 62 × 10^5 clusecs.
Note: This unit is used for the measurement of the power of evacuation of a vacuum pump.

COMFORT INDEX, ACOUSTICAL   ACI (O)

Quantity: Noise inside an aircraft cabin (none).
Definition: The ACI is a value on an arbitrarily designed scale in which:
+ 100 corresponds to zero noise
0 corresponds to conditions that are just tolerable
− 100 corresponds to intolerable conditions

COULOMB   C

Quantity: Electric charge (SI).
Definition: One coulomb is the charge transported in one second (s) by a current of one ampère (A). 1 C = 1 A s.
Note: This unit was formerly called the absolute coulomb (C_{abs}), and must be distinguished from the international coulomb (C_{int}), formally abandoned in 1948: 1 C_{int} = 1·000 34/1·000 49 C ≈ 0·999 850 C. 1 C ≈ 1·000 15 C_{int}.

COULOMB, THERMAL

Quantity: Thermal charge (SI).
Definition: One thermal coulomb corresponds to an increase in entropy of one joule per kelvin (J/K). 1 thermal coulomb = 1 J/K.
Note: The former definition of the thermal coulomb was such that it corresponded to a quantity of heat of one joule.

CRINAL

Quantity: Force (metric).
Definition: 1 crinal = 0·1 newton (N). 1 N = 10 crinals.

CRITH

Quantity: Mass, particularly the mass of a gas (arbitrary).
Definition: One crith is the mass of one litre of hydrogen at standard temperature and pressure. The experimentally derived value is:
1 crith = 8·988 5 × 10^{-5} kg. 1 kg ≈ 11 125 criths.
Note: The mass of one litre of any gas at standard temperature and pressure measured in criths is numerically equal to one-half of its relative molecular mass (molecular weight).

CROCODILE

Quantity: Electric potential and potential difference, electromotive force (metric).
Definition: 1 crocodile = 10^6 volts (V). 1 V = 10^{-6} crocodile.
Note: This unit is employed at an informal level in a number of UK nuclear physics laboratories.

CRON

Quantity: Time (arbitrary).
Definition: 1 cron = 10^6 years ≈ 3·156 × 10^{13} seconds (s).
1 s ≈ 3·169 × 10^{-14} cron.
Note: This unit has almost never been employed.

C-SIZE

Quantity: A size to which (trimmed) paper and board is manufactured (metric).
Definition: The criteria for forming members of the series are:
(1) If, for any member of the series, x = the length of the shorter sides, and y = the length of the longer sides, the next member is obtained by dividing the previous one into two along the perpendicular bisector of its longer sides. Thus for the next member, length of the longer sides = x, and length of the shorter sides = $\frac{1}{2}y$. Since the shapes of the two members are similar,

$$y : x = x : \tfrac{1}{2}y.$$
Thus $\qquad\qquad y : x = \sqrt{2} : 1.$

(2) The basic member of the series has an area of $2^{\frac{1}{4}}$ square metres (m^2). Thus, $xy = 2^{\frac{1}{4}}$ m^2. This gives $x = 2^{-1/8}$ m and $y = 2^{3/8}$ m. These values are conventionally taken as $x = 0·917$ m and $y = 1·297$ m.

Note: (1) When dividing odd numbers by 2 the result is rounded down to the nearest whole number.

(2) The manufacturing tolerances allowed are $\pm 1 \cdot 5$ mm for dimensions up to and including 150 mm, ± 2 mm for dimensions greater than 150 mm and up to and including 600 mm, and ± 3 mm for dimensions greater than 600 mm.

(3) Alternative metric series are the A-, B-, RA- and SRA-sizes.

(4) The *x*- and *y*-values of C-sizes are the geometric means of the corresponding A- and B-size values.

Members of the C-series

designation	size x × y mm mm		imperial equivalent x × y in in	
C0	917	1 297	36·10	51·07
C1	648	917	25·53	36·10
C2	458	648	18·03	25·53
C3	324	458	12·72	18·03
C4	229	324	9·02	12·72
C5	162	229	6·36	9·02
C6	114	162	4·49	6·36
C7	81	114	3·18	4·49
C8	57	81	2·24	3·18
C9	40	57	1·57	2·24
C10	28	40	1·10	1·57

CUMEC

Quantity: Volume flow rate (SI).

Definition: One cumec is the rate at which a volume of one cubic metre (m^3) flows in one second (s). 1 cumec $= 1 \ m^3/s$.

CURIE

Quantity: Radioactive disintegration rate, ie activity (arbitrary).

Symbol: Ci (BS); formerly c (BS).

Definition: The standardised value is the quantity of a radioactive nuclide required to produce $3 \cdot 7 \times 10^{10}$ disintegrating atoms per second, or, simply, $3 \cdot 7 \times 10^{10}$ disintegrations per second. 1 Ci $= 3 \cdot 7 \times 10^4$ rutherfords (rd). 1 rd $\approx 2 \cdot 702 \ 70 \times 10^{-5}$ Ci.

Note: The original definition is: One curie is the quantity of radon that is in radioactive equilibrium with one gramme of radium. The quantity of radon involved is $0 \cdot 66 \ mm^3$ approximately at standard temperature and pressure, and this gives rise to 3.61×10^{10} disintegrating atoms per second approximately. The definition was later modified to: One curie is the quantity of a radioactive nuclide producing the same disintegration rate as one gramme of radium. The present definition superceded this one.

CUSEC

Quantity: Volume flow rate (FPS).
Definition: One cusec is the rate at which a volume of one cubic foot (ft^3) flows in one second (s). 1 cusec = 1 ft^3/s. Since 1 ft = 0.304 8 m, 1 cusec = $0.304\ 8^3$ m^3/s = 0.028 316 846 592 m^3/s. 1 m^3/s $\approx$ 35.314 7 cusecs.

CYCLE PER SECOND

Quantity: Frequency (all).
Symbol: c/s; the unit symbol cps is deprecated.
Definition: One cycle per second is the frequency of a periodic occurrence which has a period of one second (s). 1 c/s = 1 /s.
Note: This unit, sometimes (incorrectly) referred to in short as the cycle, is better called the hertz. It has also been called the vibration.

DALTON

Quantity: Mass (arbitrary).
Definition: One dalton is equal to one-twelfth of the mass of a neutral carbon-12 atom. The experimentally derived value is:
1 dalton = $(1.660\ 33 \pm 0.000\ 05) \times 10^{-27}$ kg. 1 kg $\approx 6.022\ 90 \times 10^{26}$ daltons.
Note: This unit is also called the atomic mass unit (international).

DANIELL

Quantity: Potential and potential difference, electromotive force (arbitrary).
Definition: 1 daniell = 1.042 volts.
Note: This is an obsolete unit of voltage. It was meant to be the electromotive force of a Daniell cell, although this is now known to have a value of 1.08 volts.

DARAF

Quantity: Elastance, ie reciprocal capacitance (SI).
Definition: One daraf is the elastance of a substance which has a capacitance of one farad (F). 1 daraf = 1 /F.
Note: This unit is rarely employed; the reciprocal farad is used instead.

DARWIN

Quantity: Evolutionary rate of change (none).
Definition: Consider one dimension of part of an animal or plant, or of the whole animal or plant (eg its height). If, as a result of evolution, that dimension increases from s_0 to s_t (expressed in the same units) in a time t (in years) according to the formula $s_t = s_0 e^{Et/10^6}$, its evolutionary rate of change is E and is measured in darwins.

DAY d

Quantity: Time (arbitrary).
Definition: 1 d = 24 hours = 86 400 s. 1 s $\approx$ 1·157 41 $\times$ 10^{-5} d.
Note: This is the definition of the mean solar day, the time interval between consecutive passages of the sun across the meridian, averaged over one year. It must be distinguished from the sidereal day, the time for one complete rotation of the earth on its axis; the experimentally observed size is 1 sidereal day = 23 h 56 m 4·098 92 s = 86 164·098 92 s.

DEBYE D (O)

Quantity: Electric dipole moment (metric).
Definition: 1 D = 10^{-18} franklin centimetre $\approx$ 3·335 64 $\times$ 10^{-30} coulomb metre (C m). 1 C m = 2·997 925 $\times$ 10^{29} D.

DECA-...

See Appendix 3 section 2. The only unit referred to in Part 1 is:
  decatorr: see centihg (note)

DECI-...

See Appendix 3 section 2. The only units referred to in Part 1 are the following:
  decibel: see bel (note 1)
  decibel, perceived noise (entered in its alphabetical position)

DECIBEL, PERCEIVED NOISE PNdB (BS)

Quantity: Perceived noise level (all).
Definition: The perceived noise level in PNdB is equal to the sound pressure level in decibels (above a datum level of 2 $\times$ 10^{-5} pascals root-mean-square), as judged by an otologically normal binaural listener, of a band of random noise of width one-third to one octave centred on a frequency of one thousand hertz.
Note: The unit is related directly only to the noy. It is analogous in meaning to the phon.

DECILIT

Quantity: Intensity level (all).
Definition: 1 decilit = 1 decibel.
Note: This was one of several names proposed as alternatives for the decibel, but almost never employed.

DECILOG

Quantity: Intensity level (all).
Definition: 1 decilog = 1 decibel.
No   This was one of several names proposed as alternatives for the decibel, but almost never employed.

DECILU

Quantity: Intensity level (all).
Definition: 1 decilu = 1 decibel.
Note: This was one of several names proposed as alternatives for the decibel, but almost never employed.

DECOMLOG

Quantity: Intensity level (all).
Definition: 1 decomlog = 1 decibel.
Note: This was one of several names proposed as alternatives for the decibel, but almost never employed.

DEGREE (of arc)

Quantity: (Geometrical) plane angle (all).
Symbol: °. The unit symbol deg, which has been used, should be reserved for the unit of temperature difference.
Definition: $1° = \dfrac{1}{90}\llcorner = \dfrac{\pi}{180}$ rad.
Note: When using decimal fractions of a degree the unit symbol is placed after the figures, eg 7·5 °. In astronomical work it generally precedes the decimal point, eg 7 °·5.

DEGREE (of hardness)

Quantity: Hardness of water (arbitrary).
Definition: There are several types of degree of hardness:
  English degree
  Clark degree (° Clark) } 1 part calcium carbonate to 70 000 parts water*

French degree   1 part calcium carbonate to 10^5 parts water
German degree   1 part calcium oxide to 10^5 parts water

(* Equivalent to 1 grain of calcium carbonate to 1 gallon of water.)
Note: The descriptive terms used are as follows:

descriptive term	hardness		in the US
	in the UK		
	° Clark	ppm	ppm
soft	0	0	0
	5	70	55
slightly hard			
	10	140	100
moderately hard			
	15	210	200
very hard			
	> 15	> 210	> 200

DEGREE (of time)

Quantity: Time (arbitrary).
Definition: 1 degree = 0·01 day = 864 seconds (s). 1 s ≈ 0·001 157 41 degree.
Note: A metric unit suggested for the measurement of time, but almost never employed. It has also been called the cé.

DEGREE ABSOLUTE

See kelvin (1) (note 3)

DEGREE API

Quantity: Relative density (none).
Definition: The relative density d of a liquid in degrees API is related to the density relative to water S60/60, both liquids being at 60 °F, by the defining equation $d = \dfrac{141 \cdot 5}{S60/60} - 131 \cdot 5$.

Note: API is an abbreviation of American Petroleum Industry.

DEGREE BAUMÉ   °B (O)

Quantity: Relative density (none).
Definition: The relative density d of a liquid in degrees Baumé is related to

the density relative to water S15/15, both liquids being at 15 °C, by the defining equations:

(1) for relative densities less than 1, $\quad d = \dfrac{144\cdot3}{S15/15} - 144\cdot3;$

(2) for relative densities greater than 1, $\quad d = 144\cdot3 - \dfrac{144\cdot3}{S15/15}.$

DEGREE CELSIUS °C

Quantity: Customary temperature (metric).
Definition: The temperature θ_C in °C is related to the corresponding temperature T_K in kelvins (K) by the formula:
$\theta_C = T_K - 273\cdot15.$ $T_K = \theta_C + 273\cdot15.$ Thus 0 °C $= 273\cdot15$ K.
Note: (1) This definition is true for temperatures on the thermodynamic scale and the international scale of 1968; there are slight differences between the thermodynamic and practical scales.
(2) The freezing point of pure air-free water is 0 °C and its boiling point is 100 °C (both under a pressure of one (standard) atmosphere). The original Celsius scale had these values reversed.
(3) The unit was formerly the degree Centigrade. The use of this name is deprecated, although it is still employed by meteorologists in the UK. The use of the term centigrade should be restricted to the unit 0·01 grade.
(4) See also the entry Celsius degree.

DEGREE CENTIGRADE

See degree Celsius (note 3)

DEGREE FAHRENHEIT °F

Quantity: Customary temperature (arbitrary).
Definition: The temperature θ_F in °F is related to the corresponding temperature θ_C in degrees Celsius by the formula $\theta_F = \dfrac{9}{5}\theta_C + 32.$ Since θ_C is related to the corresponding temperature T_K in kelvins (K) by the formula
$\theta_C = T_K - 273\cdot15,$ $\theta_F = \dfrac{9}{5}T_K - 459\cdot67.$ $T_K = \dfrac{5}{9}(\theta_F + 459\cdot67).$
Thus 0 °F $\approx 255\cdot37$ K.
Note: (1) The freezing point of pure air-free water is 32 °F and its boiling point is 212 °F (both under a pressure of one (standard) atmosphere).
(2) See also the entry Fahrenheit degree.

DEGREE KELVIN °K

Quantity: Absolute, ie thermodynamic temperature (SI).
Definition: 1 °K = 1 kelvin (1st definition).
Note: (1) This is the former name of the unit now called the kelvin.
(2) See also the entry Kelvin degree.

DEGREE RANKINE °R

Quantity: Absolute, ie thermodynamic temperature (arbitrary).
Definition: The temperature T_R in °R is related to the corresponding temperature θ_F in degrees Fahrenheit by the formula $T_R = \theta_F + 459{\cdot}67$. Since θ_F is related to the corresponding temperature T_K in kelvins (K) by the formula $\theta_F = \frac{9}{5}T_K - 459{\cdot}67$, $T_R = \frac{9}{5}T_K$. $T_K = \frac{5}{9}T_R$. Thus 0 °R = 0 K.
Note: (1) The freezing point of pure air-free water is 491·67 °R and its boiling point is 671·67 °R (both under a pressure of one (standard) atmosphere).
(2) See also the entry Rankine degree.

DEGREE RÉAUMUR °r (O)

Quantity: Customary temperature (arbitrary).
Definition: The temperature θ_r in °r is related to the corresponding temperature θ_C in degrees Celsius by the formula $\theta_r = \frac{4}{5}\theta_C$. Since θ_C is related to the corresponding temperature T_K in kelvins (K) by the formula $\theta_C = T_K - 273{\cdot}15$, $\theta_r = \frac{4}{5}T_K - 218{\cdot}52$. $T_K = \frac{5}{4}\theta_r + 273{\cdot}15$. Thus 0 °r = 273·15 K.
Note: (1) The freezing point of pure air-free water is 0 °r and its boiling point is 80 °r (both under a pressure of one (standard) atmosphere).
(2) This unit is obsolete and has been abandoned.
(3) See also the entry Réaumur degree.

DEGREE SIKES

Quantity: Concentration (arbitrary).
Definition: This is an arbitrary measurement which, used in conjunction with a set of tables, gives the concentration of an alcohol/water mixture:

0 degrees Sikes corresponds to	66·7 over proof
10	58·4
100	pure water

Note: Proof spirit is spirit (ie alcohol/water mixture) with a density 12/13 of that of pure distilled water, both liquids at 51 °F. Its relative density (60 °F/60 °F) is 0·919 76, there being at this temperature 49·28 % alcohol by mass (57·10 % alcohol by volume). On dilution, 100 volumes of spirit of strength *P* over proof yields (100 + *P*) volumes of proof spirit.

DEGREE, SQUARE

Quantity: (Geometrical) solid angle (all).
Definition: 1 square degree = $(\pi/180)^2$ steradian (sr) $\approx 3\cdot046\ 17 \times 10^{-4}$ sr. 1 sr $\approx 3\ 282\cdot81$ square degrees.

DEGREE TWADDELL   °Tw (O)

Quantity: Relative density (none).
Definition: The relative density *d* of a liquid in degrees Twaddell is related to the density relative to water S60/60, both liquids being at 60 °F, by the defining equation $d = 200\ (S60/60 - 1)$.

DEMAL

Quantity: Concentration (metric-derived).
Definition: One demal is a concentration of one gramme-equivalent (g-eq) of solute in one cubic decimetre (dm³) of solvent.
1 demal = 1 g-eq/dm³ = 1 000 g-eq/m³ = 1 kg-eq/m³.

DENIER

Quantity: Line density (metric).
Definition: One denier is the line density of a thread which has a mass of one gramme (g) and a length of 9 000 metres (m).

1 denier = $\dfrac{1}{9\ 000}$ g/m $\approx 1\cdot111\ 11 \times 10^{-7}$ kg/m. 1 kg/m = 9×10^6 deniers.

Note: This unit is used in the textile industry as a measure of yarn count. The unit the tex is used in preference to the denier.

DEX

Quantity: Logarithmic value (all).
Definition: For a value 10^x, the size of the value is given as *x* dex. *x* dex corresponds to antilog *x*.
Note: (1) The dex may be related to the decibel (dB), the neper (Np) and the octave: 1 dex = 10 dB, 1 dB = 0·1 dex; 1 dex = ln 10 Np $\approx 2\cdot302\ 59$ Np, 1 Np $\approx 0\cdot434\ 294$ dex; 1 dex = (1/log 2) octave $\approx 3\cdot321\ 93$ octave, 1 octave $\approx 0\cdot301\ 030$ dex.
(2) The unit is also called the brig.

DIOPTRE

Quantity: Reciprocal length, especially the power of a lens (SI).
Definition: One dioptre is the reciprocal length of a distance which has a length of one metre (m). In particular, it is the power of a lens which has a focal length of one metre. 1 dioptre = 1 /m. 1 m = 1 /dioptre.
Note: (1) According to the convention known familiarly as 'real-is-positive', the power of a converging lens has a positive value and that of a diverging lens a negative value.
(2) In the US the name of this unit is spelled diopter.

DOLLAR

Quantity: Reactivity (a measure of the departure of a nuclear reactor from its critical condition) (all).
Definition: One dollar is the amount of reactivity equal to the delayed neutron fraction.
Note: (1) The delayed neutron fraction is the ratio of the mean number of delayed neutrons per fission to the mean total number of neutrons per fission.
(2) The unit is rarely used.

DONKEY POWER

Quantity: Power (metric).
Definition: 1 donkey power = 250 watts (W). 1 W = 0·004 donkey power.
Note: 1 donkey power = $\frac{1}{3}$ horsepower approximately.

DRACHM

Quantity: Mass (imperial).
Symbol: In the UK there is no unit symbol. In the US, where the unit is spelled dram and often called by the fuller name apothecaries' dram, the unit symbol is dram ap. The sign ℥ was used in early times.
Definition: 1 drachm = $\frac{1}{8}$ apothecaries' ounce = 60 grains = 0·003 887 934 6 kg. 1 kg ≈ 257·206 drachms.
Note: This is a unit of apothecaries' measure.

DRACHM, FLUID fl dr (BS)

Quantity: Capacity, ie volume.
Definition: (1) UK unit (arbitrary)
    1 UK fl dr = $\frac{1}{8}$ UK fluid ounce = $\frac{1}{160}$ UK pint ≈ 3·551 63 × 10⁻⁶ m³.
    1 m³ ≈ 2·815 61 × 10⁵ UK fl dr.
(2) US unit (imperial)
    1 US fl dr = $\frac{1}{8}$ US fluid ounce = $\frac{1}{128}$ US liquid pint ≈ 3·696 69 × 10⁻⁶ m³.
    1 m³ ≈ 2·705 12 × 10⁵ US fl dr.

Note: The UK fluid drachm is used for the measurement of liquid and solid substances, although usually the former; the US unit, spelled fluid dram, is used only for the measurement of liquid substances.

1 UK fl dr $\approx$ 0·960 759 US fl dr; 1 US fl dr $\approx$ 1·040 84 UK fl dr

DRAM   dr (BS)

Quantity: Mass (imperial).
Definition: 1 dr $= \frac{1}{16}$ ounce (avoirdupois) $= \frac{1}{256}$ lb $\approx$ 0·001 771 85 kg.
1 kg $\approx$ 564·383 dr.
Note: This is an obsolete UK avoirdupois unit; there is no corresponding US unit. The US unit called the dram is the same as the UK unit called the drachm.

DREX

Quantity: Line density (metric).
Definition: One drex is the line density of the thread which has a mass of one gramme (g) and a length of ten kilometres (km).
1 drex $= 0·1$ g/km $= 10^{-7}$ kg/m. 1 kg/m $= 10^{7}$ drex.
Note: This unit is used in the textile industry as a measure of yarn count. The unit the tex is used in preference to the drex.

D-UNIT

Quantity: Radiation dose due to X-rays (arbitrary).
Definition: 1 D-unit = 102 röntgens (R). 1 R $\approx$ 0·009 803 92 D-unit.
Note: This unit is obsolete.

DUTY

Quantity: Energy (FlbfS).
Definition: One duty is the work done when the point of application of a force of one pound-force (lbf) is displaced through a distance of one foot (ft) in the direction of the force.
1 duty $= 1$ ft lbf $= 1·355$ 817 948 331 400 4 joule (J). 1 J $\approx$ 0·737 562 duty.
Note: This is an obsolete unit. The foot pound-force should be used in its place.

DYNE   dyn

Quantity: Force (CGS).
Definition: One dyne is the force which, when applied to a body of mass one gramme (g), gives it an acceleration of one centimetre per second squared (cm/s²). 1 dyn $= 1$ g cm/s² $= 10^{-5}$ newton (N). 1 N $= 10^{5}$ dyn.

DYNE, LARGE

Quantity: Force (SI).
Definition: 1 large dyne = 1 newton.
Note: This was a name at one time applied to the unit now called the newton.

EINSTEIN UNIT

Quantity: Photoenergy (arbitrary).
Definition: The photoenergy E in Einstein units is related to the frequency f of the associated electromagnetic radiation in hertz by the formula $E = N_0 h f$, where N_0 = Avogadro's constant in atoms per mole (atom/mol) and h = Planck's constant in joule seconds (J s).
Note: The experimentally derived sizes of the constants are $N_0 = 6 \cdot 022\,52 \times 10^{23}$ atom/mol (international scale of atomic masses); $h = 6 \cdot 624\,9 \times 10^{-34}$ J s. $N_0 h \approx 3 \cdot 989\,9 \times 10^{-10}$ J s atom/mol.

ELECTRONVOLT   eV

Quantity: Energy, particularly in atomic studies (arbitrary).
Definition: One electronvolt is the energy acquired by an electron in passing through a potential difference of one volt. The experimentally derived size is 1 eV = $(1 \cdot 602\,10 \pm 0 \cdot 000\,07) \times 10^{-19}$ joule (J). 1 J $\approx 6 \cdot 241\,81 \times 10^{18}$ eV.
Note: This unit was originally called the equivalent volt.

EMAN

Quantity: Radioactive concentration (arbitrary).
Definition: One eman is a concentration of 10^{-7} curie (Ci) of radioactive material in one cubic metre (m³) of a medium.
1 eman = 10^{-7} Ci/m³. 1 Ci/m³ = 10^7 emans.
Note: This unit is now obsolete.

ENERGY UNIT

Quantity: Radiation dose (arbitrary).
Definition: One energy unit is the dose of radiation from which, through its associated ionising particles, the same energy is dissipated per gramme of material as is dissipated from one röntgen of hard X- or radium gamma radiation in one gramme of water. The experimentally derived value is 1 energy unit = $0 \cdot 009\,3$ joule per kilogramme (J/kg). 1 J/kg ≈ 108 energy units.
Note: This is an obsolete unit. It should not be confused with the e-unit or the E-unit.

EÖTVÖS   E (O)

Quantity: Horizontal gradient of gravitational acceleration (metric).
Definition: One eötvös is a change in gravitational acceleration of 10^{-9} galileo (Gal) over a horizontal distance of one centimetre (cm).
1 E = 10^{-9} Gal/horizontal centimetre = 10^{-9} (m/s^2)/horizontal m.
1 (m/s^2)/horizontal m = 10^9 E.

ERG

Quantity: Energy (CGS).
Definition: One erg is the work done when the point of application of a force of one dyne (dyn) is displaced through a distance of one centimetre (cm) in the direction of the force.
1 erg = 1 dyn cm = 10^{-7} joule (J). 1 J = 10^7 erg.

ERGON

Quantity: Energy associated with electromagnetic radiation (arbitrary).
Definition: The energy E in ergons is related to the frequency f of the electromagnetic radiation in hertz by the formula $E = hf$, where h = Planck's constant in joule seconds (J s).
Note: (1) The experimentally derived size of h is $h = 6.624\,9 \times 10^{-34}$ J s.
(2) The unit is better called the quantum; it is also called the photon.

ERLANG   E (BS)

Quantity: Telephone traffic intensity (any).
Definition: One erlang is the telephone traffic intensity resulting from a call rate of one call per hour, the call taking one hour to make.
Note: The telephone traffic intensity is the product of the number of calls made in a given period of time and the average length of the calls (measured in the same time unit).

e-UNIT

Quantity: Radiation dose due to X-rays (arbitrary).
Definition: There is no formal definition. 1 e-unit = 7 röntgens approximately.
Note: This unit is obsolete. It must not be confused with the E-unit or the energy unit.

E-UNIT

Quantity: Radiation dose rate, ie intensity, due to X-rays (arbitrary).

Definition: There is no formal definition. 1 E-unit = 1 röntgen per second approximately.
Note: This unit is obsolete. It must not be confused with the e-unit or the energy unit.

FAHRENHEIT DEGREE   degF

Quantity: Temperature interval or difference (arbitrary).
Definition: 1 degF = $\frac{5}{9}$ kelvin (K). 1 K = 1·8 degF.
Note: (1) The original definition was: 1 degF is $\frac{1}{180}$ of the interval between the freezing and boiling points of pure air-free water, both under a pressure of one (standard) atmosphere.
(2) See also the entry degree Fahrenheit.

FARAD   F

Quantity: Electric capacitance (SI).
Definition: One farad is the capacitance of a condenser having a charge of one coulomb (C), across the plates of which the potential difference is one volt (V). 1 F = 1 C/V.
Note: (1) This unit was formerly called the absolute farad (F_{abs}), and must be distinguished from the international farad (F_{int}), formally abandoned in 1948: 1 F_{int} = 1/1·000 49 F ≈ 0·999 510 F. 1 F = 1·000 49 F_{int}.
(2) The original definition of the farad gave it a size equal to one present-day microfarad.

FARADAY

Quantity: Charge (arbitrary).
Definition: One faraday is the charge necessary to liberate one gramme-equivalent of a substance in electrolysis. The experimentally derived value is 1 faraday = 96 487·0 ± 1·6 coulombs (C). 1 C ≈ 1·036 41 × 10^{-5} faraday. (These values are on the international scale of atomic masses.)

FARAD, THERMAL

Quantity: Thermal capacitance (SI).
Definition: One thermal farad is the thermal capacitance for which an amount of entropy of one joule per kelvin (J/K) added to a body raises its temperature by one kelvin. 1 thermal farad = 1 J/K^2.
Note: The former definition of the thermal farad was such that it corresponded to a quantity of heat of one joule resulting in a temperature increase of one kelvin. 1 thermal farad = 1 J/K.

FATHOM

Quantity: Length, particularly marine depth (imperial).
Definition: 1 fathom = 6 ft = 1·828 8 m. 1 m $\approx$ 0·546 807 fathom.
Note: See Appendix 4 for (cubic) fathom, (square) fathom.

FEMTO-. . .

See Appendix 3 section 2. Units are not entered under the prefix femto-.

FERMI

Quantity: Length, especially in atomic studies (metric).
Definition: 1 fermi = 10^{-15} m. 1 m = 10^{15} fermis.
Note: The unit is no longer used.

FINSEN UNIT FU (O)

Quantity: Intensity of ultraviolet radiation (metric).
Definition: One finsen unit is the intensity of ultraviolet radiation of a specified wavelength with an energy density of 100 000 watts per metre squared (W/m²).
1 FU $\equiv$ 10^5 W/m². 1 W/m² $\equiv$ 10^{-5} FU.
Note: The wavelength normally specified is 296·7 nm.

FLUX UNIT fu (O)

Quantity: Flux density of radio-astronomical sources (metric).
Definition: 1 fu = 10^{-26} watt per metre squared hertz (W/m² Hz).
1 W/m² Hz = 10^{26} fu.

FOOT

Quantity: Length (FPS, FSS, FlbfS).
Symbol: ft. The sign ' is sometimes used, but this is not recommended.
Definition: 1 ft = $\frac{1}{3}$ yard = 0·304 8 m. 1 m $\approx$ 3·280 84 ft.
Note: See Appendix 4 for board foot.

FOOT-CANDLE fc (O)

Quantity: (Intensity of) illumination (FPS).
Definition: One foot-candle is the illumination of one lumen (lm) uniformly over an area of one square foot (ft²).
1 fc = 1 lm/ft² $\approx$ 10·763 9 lux (lx). 1 lx = 0·092 903 04 fc.
Note: The name foot-candle is deprecated; it should be replaced by lm/ft².
The name candle-foot was used originally.

FOOT-CANDLE, EQUIVALENT

Quantity: Luminance (imperial).
Definition: One equivalent foot-candle is the luminance of a uniform diffuser emitting one lumen per square foot (ft²). In terms of the candela (cd), 1 equivalent foot-candle = $1/\pi$ cd/ft² = $1/(0.304\ 8^2\ \pi)$ nit (nt) $\approx$ 3·426 25 nt. 1 nt $\approx$ 0·291 864 equivalent foot-candle.
Note: (1) The use of this unit is deprecated; the candela per foot squared should be used in its place.
(2) The unit is better called the foot-lambert.

FOOT, CAPE

Quantity: Length (imperial-derived).
Definition: 1 Cape foot = 1·033 foot = 0·314 858 4 m. 1 m $\approx$ 3·176 03 Cape feet.

FOOT-LAMBERT   ft-L

Quantity: Luminance (imperial).
Definition: One foot-lambert is the luminance of a uniform diffuser emitting one lumen per square foot (ft²). In terms of the candela (cd):
    1 ft-L = $1/\pi$ cd/ft² = $1/(0.304\ 8^2\pi)$ nit (nt) $\approx$ 3·426 25 nt.
    1 nt $\approx$ 0·291 864 ft-L.
Note: (1) The use of this unit is deprecated; the candela per foot squared should be used in its place.
(2) The unit has also been called the equivalent foot-candle.

FOOT, SURVEY (US)

Quantity: Length (arbitrary).
Definition: 1 US survey foot = 12/39·37 m $\approx$ 1·000 002 ft $\approx$ 0·304 800 6 m. 1 m $\approx$ 3·280 83 US survey feet.
Note: This unit is used by the US Coast and Geodetic Survey.

FORS (1)   f (O)

Quantity: Acceleration; specific force, ie force per unit mass (metric).
Definition: One fors is equal to the standard acceleration of free fall. 1 f = 9·806 65 m/s². 1 m/s² $\approx$ 0·101 972 f.
Note: The unit is better called the G. It has also been called the grav.

FORS (2)   f (O)

Quantity: Force (metric).
Definition: One fors is the force which, when applied to a body of mass one

gramme, gives it an acceleration equal to the local value of the acceleration of free fall (g) expressed in centimetres per second squared.

1 f = g dyne = g × 10^{-5} newton (N). 1 N = $10^5/g$ f.

Note: This is an inconsistent unit, better known as the gramme-weight, and its use is deprecated. The gramme-force should always be used in its place.

FOURIER

Quantity: Thermal resistance (SI).

Definition: One fourier is the thermal resistance for which a temperature difference of one kelvin (K) causes an entropy flow of one watt (W) per kelvin. 1 fourier = 1 K^2/W.

Note: (1) Formerly, one fourier was the thermal resistance that allows a heat flow rate of one watt under a temperature difference of one kelvin. 1 fourier = 1 K/W.

(2) The unit is better called the thermal ohm.

FRANKLIN Fr

Quantity: Charge (CGSe).

Definition: One franklin is that charge which exerts on an equal charge at a distance of one centimetre in a vacuum a force of one dyne. 1 Fr = $0\cdot1/(c)$ coulomb (C), where (c) is the numerical value of the velocity of electromagnetic radiation in vacuo, expressed in metres per second. 1 Fr ≈ $3\cdot335\,64 \times 10^{-10}$ C. 1 C = $2\cdot997\,925 \times 10^9$ Fr. The CGS-esu mechanical equivalent is 1 Fr = 1 $dyne^{\frac{1}{2}}$ cm. The CGS-esu equivalent is 1 Fr = 1 statcoulomb (statC).

FRAUNHOFER F (0)

Quantity: Reduced width of a spectrum line (all).

Definition: The reduced width W of a spectrum line in fraunhofers is related to its wavelength λ and equivalent width Δλ, both measured in the same units, by the formula $W = 10^6\,\Delta\lambda/\lambda$.

FRESNEL

Quantity: Frequency (all).

Definition: 1 fresnel = 10^{12} hertz (Hz). 1 Hz = 10^{-12} fresnel.

FRIGORIE fg

Quantity: Heat energy (arbitrary).

Definition: One frigorie corresponds to the extraction of one thousand calories (cal_{15}) of heat from the body to be cooled. From the experimentally

derived value of the cal_{15}, 1 fg = 1 000 cal_{15} = 4 185·5 ± 0·5 joules (J). 1 J ≈ 2·389 2 × 10^{-4} fg.
Note: The unit is used in refrigeration engineering.

FUNAL

Quantity: Force (MTS).
Definition: One funal is the force which, when applied to a body of mass one tonne (t), gives it an acceleration of one metre per second squared (m/s²). 1 funal = 1 t m/s² ⊏ 1 000 newtons (N). 1 N = 0·001 funal.
Note: This is a former name for the unit now called the sthène.

FURLONG   fur (O)

Quantity: Length (imperial).
Definition: 1 fur = $\frac{1}{8}$ mile = 660 feet = 201·168 m. 1 m ≈ 4·970 97 × 10^{-3} fur.

G

Quantity: Acceleration.
(1) Metric unit.
   Definition: One G is equal to the standard acceleration of free fall. 1 G = 9·806 65 m/s². 1 m/s² ≈ 0·101 972 G.
   Note: The unit is sometimes used, for convenience, in aeronautical and astronautical applications. It is also called the grav or the fors.
(2) Imperial unit.
   Definition: One G is equal to the standard acceleration of free fall expressed in feet per second squared to as many significant figures as required, usually three (1 G = 32·2 ft/s²) or six (1 G = 32·174 0 ft/s²).
   Note: The unit is sometimes used, for convenience, in aeronautical and astronautical applications. It is an approximation to the metric unit: exactly, 1 G = $\dfrac{9\cdot806\ 65}{0\cdot304\ 8}$ ft/s².

GALILEO = GAL

Quantity: Acceleration (CGS).
Symbol: Gal. The unit symbol is written with a capital letter to distinguish it from that for gallon, gal.
Definition: 1 Gal = 1 cm/s² = 0·01 m/s². 1 m/s² = 100 Gal.
Note: The unit is commonly used in geodetic measurement: generally the milligal is employed, where 1 mGal = 0·001 Gal = 10^{-5} m/s².

GALLON   gal

Quantity: Capacity, ie volume.
(1) UK unit (arbitrary)
   Definition: One UK gallon is the space occupied by ten pounds weight
   of distilled water of density 0·998 859 grammes per millilitre in air of
   density 0·001 217 grammes per millilitre against weights of density 8·136
   grammes per millilitre (Weights and Measures Act of 1963).
   1 UKgal = 8 UK pints $\approx$ 4·546 09 $\times$ 10^{-3} m^3. 1 m^3 $\approx$ 219·969 UKgal.
   Note: (1) The UK gallon is used for the measurement of liquid and
   solid substances, although usually the former. The original definition,
   given in the Weights and Measures Act of 1878, is: One UK gallon is the
   volume occupied by a quantity of distilled water which has a mass of ten
   pounds in air against brass weights, the water and air at a temperature of
   sixty-two degrees of Fahrenheit's thermometer and the barometer
   standing at thirty inches of mercury.
   (2) The National Physical Laboratory accepts a value of 1 UKgal =
   277·420 cubic inches (in^3) to six significant figures.
(2) US unit (imperial)
   Definition: 1 US gallon = 231 in^3 = 8 US liquid pints = 3·785 411 784 $\times$
   10^{-3} m^3. 1 m^3 $\approx$ 264·172 USgal.
   Note: (3) The US gallon is used only for the measurement of liquid
   substances.
Note: (4) 1 UKgal $\approx$ 1·200 95 USgal; 1 USgal $\approx$ 0·832 674 UKgal.
(5) See Appendix 4 for another entry 'gallon', barn gallon and 'Winchester'
wine gallon.

GALVAT

Quantity: Current (MKSA).
Definition: 1 galvat = 1 international ampère.
Note: This name was proposed for the international unit of current, but
never employed.

GAMMA   γ (O) (1)

Quantity: Magnetic field strength (metric).
Definition: 1 γ = 10^{-5} oersted $\approx$ 7·957 75 $\times$ 10^{-4} ampère per metre (A/m).
1 A/m $\approx$ 1·256 64 $\times$ 10^3 γ.

GAMMA   γ (BS) (2)

Quantity: Mass (metric)
Definition: 1 γ = 1 μg = 10^{-9} kg. 1 kg = 10^9 γ.

GAMMIL

Quantity: Concentration (metric).
Definition: One gammil is a concentration of one milligramme (mg) of solute in one litre (l) of solvent. 1 gammil = 1 mg/l. Using the new litre, 1 gammil = 0·001 kilogramme per metre cubed (kg/m³). 1 kg/m³ = 1 000 gammils.
Note: This unit has also been called the microgammil and the micril. It has rarely been used.

GAUSS

Quantity: Magnetic flux density (CGS-emu).
Symbol: G in physics; Gs in engineering.
Definition: One gauss is the area-density of one maxwell (Mx) of magnetic flux per square centimetre (cm²). 1 G = 1 Mx/cm² = 10^{-4} tesla (T). 1 T = 10^4 G. The CGS-emu equivalent is 1 G = 1 abtesla. The CGS-emu mechanical equivalent is 1 G = 1 dyne$^{\frac{1}{2}}$/cm. The CGSm equivalent is 1 G = 1 erg/biot cm².
Note: The name gauss was proposed at the 3rd International Electrical Congress of 1891 for the practical unit of magnetic flux density (for which the name international tesla is now used). It was proposed by the American Institute of Electrical Engineers in 1894 for the rationalised CGS-emu of magnetic flux density (rather than unrationalised, as at present). It was proposed by the 5th International Electrical Congress of 1900 for the CGS-emu of magnetic field strength as well as flux density, these two quantities being at the time regarded as dimensionally equivalent. The IEC fixed its present use in 1930.

GEE POUND

Quantity: Mass (FSS).
Definition: One gee pound is the mass that acquires an acceleration of one foot per second squared under the influence of a force of one pound-force. 1 gee pound = 9·806 65/0·304 8 lb ≈ 32·174 0 lb ≈ 14·593 9 kg.
1 kg ≈ 0·068 521 8 gee pound.
Note: This is the British technical unit of mass. It is better called the slug.

GEMMHO

Quantity: Conductance (metric).
Definition: One gemmho is the conductance of a substance which has a resistance of one megohm (MΩ).
1 gemmho = 1 /MΩ = 10^{-6} siemens (S). 1 S = 10^6 gemmhos.
Note: This unit is rarely employed.

GIBBS

Quantity: Absorption, ie surface concentration (metric).
Definition: One gibbs is an adsorption of one micromole (10^{-6} mol) over one square metre (m^2). 1 gibbs $= 10^{-6}$ mol/m^2. 1 mol/$m^2 = 10^6$ gibbs.

GIGA-...

See Appendix 3 section 2. Units are not entered under the prefix giga-.

GILBERT

Quantity: Magnetomotive force (CGS-emu).
Symbol: Gb. The unit symbol Gi has also been used.
Definition: One gilbert is the magnetomotive force resulting from the passage of a current of 4π abampères (abA) through one turn (T) of a coil. 1 Gb $= 4\pi$ abAT $\approx 12\cdot5664$ abAT. 1 Gb $= 10/4\pi$ ampère-turn (A) $\approx 0\cdot795\,775$ A. 1 A $\approx 1\cdot256\,64$ Gb. The CGS-emu mechanical equivalent is 1 Gb $= 1$ dyne$^{\frac{1}{2}}$. The CGSm equivalent is 1 Gb $= 1$ biot (Bi).
Note: The name gilbert was proposed by the American Institute of Electrical Engineers in 1894 for the rationalised CGS-emu unit of magnetomotive force. The IEC in 1930 gave it its present (unrationalised) use.

GILL

Quantity: Capacity, ie volume.
Definition: (1) UK unit (arbitrary)
    1 UK gill $= \frac{1}{4}$ UK pint $\approx 1\cdot420\,65 \times 10^{-4}\,m^3$. 1 $m^3 \approx 7\,039\cdot03$ UK gills.
(2) US unit gi (O) (imperial)
    1 USgi $= \frac{1}{4}$ US liquid pint $= 1\cdot182\,941\,182\,5 \times 10^{-4}\,m^3$. 1 $m^3 \approx 8\,453\cdot51$
    USgi.
Note: The UK gill is used for the measurement of liquid and solid substances, although usually the former; the US gill is used only for the measurement of liquid substances. 1 UK gill $\approx 1\cdot200\,95$ USgi; 1 USgi $\approx 0\cdot832\,675$ UK gill.

GLUG

Quantity: Mass (CgfS)
Definition: One glug is the mass that acquires an acceleration of one centimetre per second squared under the influence of a force of one gramme-force. 1 glug $= 980\cdot665$ g $= 0\cdot980\,665$ kg. 1 kg $\approx 1\cdot019\,72$ glug.
Note: The name of this unit is not often used; it is generally referred to as the CGS-technical unit of mass.

GON

Quantity: (Geometrical) plane angle (all).
Definition: 1 gon = 0·01 right angle = $\pi/200$ radian = 0·9 °.
Note: This unit is better called the grade.

GRADE ^{g}

Quantity: (Geometrical) plane angle (all).
Definition: 1 ^{g} = 0·01 $\llcorner$ = $\pi/200$ rad = 0·9 °.
Note: This unit has only been used on the continent of Europe. It has also been called the gon.

GRAIN gr (BS)

Quantity: Mass (imperial).
Definition: 1 gr = $\dfrac{1}{7\,000}$lb = 6·479 891 × 10^{-5} kg. 1 kg ≈ 1·543 24 × 10^4 gr.
Note: The size of this unit is common to the avoirdupois, apothecaries' and troy systems.

GRAIN, METRIC

Quantity: Mass (metric).
Definition: 1 metric grain = 50 milligrammes = 5 × 10^{-5} kg.
1 kg = 2 × 10^4 metric grains.
Note: This unit is employed for commercial transactions in diamonds, pearls and other precious stones.

GRAMME g

Quantity: Mass (CGS).
Definition: 1 g = 0·001 kg. 1 kg = 1 000 g.
Note: (1) In the US the name of this unit is spelled gram. The spelling in official translations of ISO Recommendations is always gramme.
(2) The names bes, brieze and stathm have been proposed as alternatives for the gramme, but never employed.

GRAMME-CALORIE cal (O)

Quantity: Heat energy (arbitrary).
Definition: One gramme-calorie is the quantity of heat required to raise the temperature of one gramme of air-free water from 14·5 °C to 15·5 °C at a constant pressure of one (standard) atmosphere. The experimentally derived size is 1 cal = 4·185 5 ± 0·000 5 joules (J). 1 J ≈ 0·238 92 cal.
Note: The unit is better called the fifteen-degree calorie. It has also been called the small calorie.

GRAMME-EQUIVALENT   g-eq (O)

Quantity: Mass (metric-derived).
Definition: One gramme-equivalent is the mass of an element or radical in grammes equivalent to (ie that combines with or replaces) three grammes of tetravalent carbon-12.

GRAMME-FORCE   gf (BS)

Quantity: Force (CgfS).
Definition: One gramme-force is the force which, when applied to a body of mass one gramme, gives it an acceleration equal to the standard acceleration of free fall (g_n). Since g_n = 980·665 centimetres per second squared = 9·806 65 metres per second squared,
1 gf = 980·665 dynes = 0·009 806 65 newton (N). 1 N $\approx$ 101·972 gf.

GRAMME-MOLECULE   gmol (O)

Quantity: Amount of substance (metric).
Definition: 1 gmol = 1 mole.
Note: This was the former name for the mole.

GRAMME-RAD

Quantity: Integral absorbed ionising radiation dose (metric).
Definition: 1 gramme-rad = 10^{-5} joule (J). 1 J = 10^5 gramme-rads.
Note: The integral absorbed ionising radiation dose is the integral of the absorbed dose over a given volume.

GRAMME-RÖNTGEN

Quantity: Absorbed radioactive energy (arbitrary).
Definition: One gramme-röntgen is the energy absorbed when one röntgen (R) is delivered to one gramme (g) of air. 1 gramme-röntgen = 1 R g. The experimentally-derived value is 1 gramme-röntgen = 8·69 × 10^{-6} joule (J) approximately. 1 J $\approx$ 1·15 × 10^5 gramme-röntgens.

GRAMME-WEIGHT   gwt (BS)

Quantity: Force (metric).
Definition: One gramme-weight is the force which, when applied to a body of mass one gramme, gives it an acceleration equal to the local value of the acceleration of free fall (g) expressed in centimetres per second squared.
1 gwt = g dyne = g × 10^{-5} newton (N). 1 N = $10^5/g$ gwt.
Note: This is an inconsistent unit and its use is deprecated. The gramme-force should always be used in its place. It has also been called the fors.

GRAV G

Quantity: Acceleration (metric).
Definition: One grav is equal to the standard acceleration of free fall.
$1 \text{ G} = 9 \cdot 806 \, 65 \text{ m/s}^2$. $1 \text{ m/s}^2 \approx 0 \cdot 101 \, 972 \text{ G}$.
Note: The unit is sometimes used, for convenience, in aeronautical and astronautical applications. It is usually simply called the G, and has also been called the fors.

GRAVE

Quantity: Mass (metric)
Definition: One grave is the mass of one (original) litre of pure air-free water, ie it was the original name for the kilogramme.
Note: (1) Using the value of the (original) litre, 1 grave $= 1 \cdot 000 \, 028 \times 10^{-3} \text{ kg}$.
$1 \text{ kg} \approx 999 \cdot 972$ graves.
(2) This unit is quite obsolete.

HARDNESS NUMBER

Quantity: Hardness of materials (arbitrary).
(1) Brinell hardness number   HB
   Definition: Consider a steel sphere of diameter D (in millimetres, mm) which acts as an indenter on a surface. If it produces an indentation with a surface area A (in mm²), a mean diameter d (in mm) and a depth h (in mm) under a force F (in kilogrammes-force, kgf), the Brinell hardness number is defined by

$$\text{HB} = F/A$$
$$= \frac{2F}{\pi D[D - (D^2 - d^2)^{\frac{1}{2}}]}$$
$$= \frac{F}{\pi Dh}$$

   Note: (1) F and D are chosen such that $F/D^2 = k$ kgf/mm², where $k = 1, 5, 10$ or 30.
(2) Rockwell hardness number   HR
   Definition: Consider an indenter that produces an indentation in a surface under a force F_0 (in kgf); let the indentation increase to a depth h_1 when F_0 is increased by a force F_1 (in kgf), and let it decrease to h_2 when F_1 is removed. If $e = h_1 - h_2$, the Rockwell hardness number is defined by $\text{HR} = E - e$, where $E = 100$ for a diamond cone indenter and $E = 130$ for a steel sphere indenter.

Note: (2) There are nine Rockwell hardness scales.

scale designation	indenter	F_0 (kgf)	F_1 (kgf)	E
HRA	diamond cone	10	50	100
HRB	$\frac{1}{16}$ inch-diameter steel sphere	10	90	130
HRC	diamond cone	10	140	100
HRD	diamond cone	10	90	100
HRE	$\frac{1}{8}$ inch-diameter steel sphere	10	90	130
HRF	$\frac{1}{16}$ inch-diameter steel sphere	10	50	130
HRG	$\frac{1}{16}$ inch-diameter steel sphere	10	140	130
HRH	$\frac{1}{8}$ inch-diameter steel sphere	10	50	130
HRK	$\frac{1}{8}$ inch-diameter steel sphere	10	140	130

(3) Vickers hardness number   HV

Definition: Consider an indenter of square cross section tapering symmetrically to a point. If the indenter produces an indentation with a sloping area A (in mm²) and an (arithmetic) mean diagonal diameter d (in mm) under a force F (in kgf), the Vickers hardness number is defined by

$$HV = F/A$$
$$= 2\ F \sin\ \theta/d^2,$$

where θ is the half-angle of the taper of the indenter.

Note: (3) 2θ is generally taken as the standard value of 136 °. F is chosen with the value k kgf, where $k = 1, 2 \cdot 5, 5, 10, 20, 30, 50$ or 100.

Note: (4) See also table 55 in Appendix 6.

HARTLEY

Quantity: Information for a digital computer (none).

Definition: 1 hartley = $\log_2 10$ bits $\approx 3 \cdot 321\ 93$ bits.

HARTREE

Quantity: Energy, particularly in atomic studies (Hartree system of units).

Definition: 1 hartree = e^2/a_0, where e is the atomic unit of charge and a_0 is the atomic unit of length. Hence 1 hartree $\approx 4 \cdot 850\ 5 \times 10^{-18}$ joule (J). 1 J $\approx 2 \cdot 061\ 6 \times 10^{17}$ hartrees.

Note: This unit is more often called the atomic unit of energy.

HECTO-...

See Appendix 3 section 2. The only unit referred to in Part 1 is:

hectare: see are (note)

HEFNERKERZE   HK (BS)

Quantity: Luminous intensity (all).
Definition: 1 HK $\approx$ 0·901 international standard candles (measured comparison).
Note: The Hefnerkerze (Hefner candle) was the legally defined unit of intensity in Germany from 1884 to 1940, when it was replaced by the candela.

HEHNER NUMBER

Quantity: Concentration of fatty acids in oils (arbitrary).
Definition: A Hehner number of x corresponds to x kilogrammes of water-insoluble fatty acids in one hundred kilogrammes of an oil or fat.
Note: The unit is also called the Hehner value.

HELMHOLTZ

Quantity: Dipole moment per unit area (metric).
Definition: One helmholtz is equal to one debye (D) per ångström squared (Å^2).
1 helmholtz = 1 $D/\text{Å}^2$ = 0·01 franklin per centimetre $\approx$ 3·335 64 $\times$ 10^{-10} coulomb per metre (C/m). 1 C/m = 2·997 925 $\times$ 10^9 helmholtz.

HENRY   H

Quantity: Electric inductance (SI).
Definition: One henry is the inductance of a closed circuit which gives rise to a magnetic flux of one weber (Wb) per ampère (A). 1 H = 1 Wb/A.
Note: (1) This unit was formerly called the absolute henry (H_{abs}), and must be distinguished from the international henry (H_{int}), formally abandoned in 1948: 1 H_{int} = 1·000 49 H. 1 H $\approx$ 0·999 510 H_{int}. The international henry has also been called the quadrant and the secohm.
(2) The size of the standard henry has been experimentally determined with a Campbell apparatus to an accuracy (at the National Physical Laboratory) of 10 parts per million.

HENRY, THERMAL

Quantity: Thermal inductance (SI).
Definition: One thermal henry is the thermal inductance for which an entropy flow of one watt per kelvin (W/K) is associated with a kinetic energy of one joule (J). 1 thermal henry = 1 $J K^2/W^2$.
Note: The former definition of the thermal henry was such that it corresponded to a heat flow rate of one watt associated with a kinetic energy of one joule. 1 thermal henry = 1 J/W^2.

HERSCHEL

Quantity: Radiance (metric).
Definition: 1 herschel $= \pi$ watt per metre squared per steradian $(W/m^2\ sr)$. 1 $W/m^2\ sr = 1/\pi$ herschel.
Note: This unit has almost never been used.

HERTZ   Hz

Quantity: Frequency (all).
Definition: One hertz is the frequency of a periodic occurrence which has a period of one second (s). $1\ Hz = 1\ /s$. $1\ s = 1\ /Hz$.
Note: This unit is also called the cycle (per second). It has also been called the vibration.

HORSEPOWER   hp (BS)

Quantity: Power (imperial).
Definition: 1 hp $=$ 550 ft lbf/s.
Since 1 ft $= 0{\cdot}304\ 8$ m and 1 pound-force (lbf) $= 4{\cdot}448\ 221\ 615\ 260\ 5$ newtons, 1 hp $= 550 \times 0{\cdot}304\ 8 \times 4{\cdot}448\ 221\ 615\ 260\ 5$ watt (W) $= 745{\cdot}699\ 871\ 582\ 270\ 22$ W. 1 W $\approx 1{\cdot}341\ 02 \times 10^{-3}$ hp.
Note: This unit is sometimes called the British horsepower, to distinguish it from the metric horsepower. 1 hp $\approx 1{\cdot}013\ 87$ metric horsepower; 1 metric horsepower $\approx 0{\cdot}986\ 320$ hp.

HORSEPOWER, METRIC

Quantity: Power (metric).
Definition: 1 metric horsepower $=$ 75 metre kilogramme-force per second (m kgf/s). Since 1 kgf $= 9{\cdot}806\ 65$ newtons, 1 metric horsepower $= 75 \times 9{\cdot}806\ 65$ watt (W) $= 753{\cdot}498\ 75$ W. 1 W $\approx 0{\cdot}001\ 359\ 62$ metric horsepower.
Note: In France this unit is called the cheval vapeur (ch); in Germany it is called the Pferdestärke (PS).

HOUR   h

Quantity: Time (arbitrary).
Definition: 1 h $=$ 60 minutes $=$ 3 600 s. $1\ s \approx 2{\cdot}777\ 78 \times 10^{-4}$ h.

HUNDREDWEIGHT

Quantity: Mass (imperial).
(1) Avoirdupois measure   cwt (BS)
    Definition: 1 cwt $=$ 112 lb $= 50{\cdot}802\ 345\ 44$ kg. 1 kg $\approx 0{\cdot}019\ 684\ 1$ cwt.

Note: This is the UK name for the unit; in the US, where it is almost obsolete, it is called the long hundredweight. The US unit called the hundredweight is, in the UK, called the cental, the centner or the quintal.

(2) Troy measure   cwt tr (O)

Definition: 1 cwt tr = 100 troy pounds = 37·324 172 16 kg.
1 kg ≈ 0·026 792 3 cwt tr.

HUNDREDWEIGHT, SHORT   sh cwt

Quantity: Mass (imperial).
Definition: 1 sh cwt = 100 lb = 45·359 237 kg. 1 kg ≈ 0·022 046 2 sh cwt.
Note: This is a US unit; in the UK it is called the quintal, and has also been called the centner and the cental.

HYL

Quantity: Mass (metric-technical).
Definition: One hyl is the mass that acquires an accleration of one metre per second squared under the influence of a force of one kilogramme-force. 1 hyl = 9·806 65 kg. 1 kg ≈ 0·101 972 hyl.
Note: The name for this unit is not often used; it is generally simply referred to as the metric technical unit of mass. It is also called the metric slug. Alternative proposed (but unemployed) names are the mug and the par.

INCH

Quantity: Length (imperial).
Symbol: in. The sign ″ is sometimes used, but this is not recommended.
Definition: 1 in = $\frac{1}{12}$ foot = 0·025 4 m. 1 m ≈ 39·370 1 in.

INCH, CIRCULAR

Quantity: Area (imperial).
Definition: One circular inch is the area of a circle of diameter one inch (in). 1 circular inch = $\pi/4$ in² ≈ 0·785 398 in² ≈ 5·067 08 × 10⁻⁴ m².
1 m² ≈ 1 973·53 circular inches.

INHOUR   ih (O)

Quantity: Reactivity (a measure of the departure of a nuclear reactor from its critical condition) (arbitrary derived).
Definition: The reactivity R of a nuclear reactor in inhours is related to the reactor period T in hours (h) by the formula $R = 1/T$. Thus 1 ih is the reactivity that corresponds to a period of 1 h.

Note: (1) The reactor period is the time required (under conditions in which the neutron flux is varying exponentially) for the neutron flux to change by a factor of e.

(2) This unit, an abbreviation of inverse hour, is rarely employed.

INTERNATIONAL . . .

The original definitions of electrical units (including those of power and energy) were given in terms of experimentally determined quantities. They were originally designated practical units, and later international units, and the unit symbols given the subscript int. The international units were abandoned at the 9CGPM of 1948 in favour of the (theoretically defined) absolute units. The relationships between the international and absolute units of electromotive force (the volt) and resistance (the ohm) were formally defined at the CIPM of 1946, and the relationships between other international and absolute units may be calculated from them. These relationships are given in the notes accompanying the units involved:

ampère	ohm
coulomb	siemens
farad	tesla
henry	volt
joule	watt
mho	weber

IODINE NUMBER

Quantity: Percentage of absorbed iodine (arbitrary).

Definition: An iodine number of x corresponds to x kilogrammes of iodine absorbed by one hundred kilogrammes of a substance.

Note: The iodine number (also called the iodine value) is used as a measure of the proportion of unsaturated linkages present in a sample of an oil or fat.

JAR

Quantity: Capacitance (CGS-esu derived).

Definition: 1 jar = 1 000 statfarads = $10^8/(c)^2$ farad (F), where (c) is the numerical value of the velocity of electromagnetic radiation in vacuo expressed in metres per second.

Hence 1 jar $\approx$ 1·112 65 $\times$ 10^{-9} F. 1 F $\approx$ 8·987 55 $\times$ 10^8 jars.

Note: (1) This is an obsolete unit representing the approximate capacitance of a Leyden jar.

(2) See also the entry in Appendix 4.

JERK

Quantity: Rate of change of acceleration (FPS).
Definition: One jerk is a rate of change of acceleration of one foot per second squared per second. 1 jerk = 1 ft/s^3.
Note: The unit has been used by engineers in the UK.

JOULE   J

Quantity: Energy (SI).
Definition: One joule is the work done when the point of application of a force of one newton (N) is displaced through a distance of one metre (m) in the direction of the force. 1 J = 1 N m.
Note: (1) The 9CGPM of 1948 adopted this unit for the measurement of electrical energy and heat energy as well as of mechanical energy. Up to this time electrical energy was measured in units of the same size but called absolute joules (J_{abs}), and heat energy in calories. The J_{abs} must be distinguished from the (mean) international joule (J_{int}), formally abandoned in 1948: 1 J_{int} = (1·000 34)^2/1·000 49 J $\approx$ 1·000 19 J.   1 J $\approx$ 0·999 810 J_{int}.
(2) The former definition of the thermal coulomb made it equal to the joule.

k

Quantity: Reactivity (a measure of the departure of a nuclear reactor from its critical condition) (all).
Definition: 1 k is a change in reactivity of unity.
Note: (1) The name of the unit, which is rarely employed, is itself the unit symbol.
(2) The basic unit is never used, but the sub-multiple the milli-k (0·001 k).

KANNE

Quantity: Capacity, ie volume (metric).
Definition: 1 kanne = 1 litre.
Note: This name was proposed as an alternative to the litre, but never employed.

KAPP LINE (OF MAGNETIC FORCE)

Quantity: Magnetic flux (CGS-emu derived).
Definition: 1 kapp line = 6 000 maxwells = 6 $\times$ 10^{-5} weber (Wb). 1 Wb $\approx$ 16 666·7 kapp lines.
Note: This unit has only been used by its eponymous inventor.

KAYSER   K (O)

Quantity: Reciprocal length, especially wave number (CGS).
Definition: One kayser is the reciprocal length of a distance which has a length of one centimetre (cm). 1 K = 1 /cm = 100 /m. 1 m = 100 /K.
Note: This unit has also been called the balmer and the rydberg.

KELVIN   (1), (2)

(1) *Quantity:* Absolute, ie thermodynamic temperature (SI).
   Symbol: K; it was formerly °K.
   Definition: The official definition of this basic SI unit, given at the 10CGPM of 1954, is: La dixième Conférence Générale des Poids et Mesures décide de définir l'échelle thermodynamique de température au moyen du point triple de l'eau comme point fixe fondamental, en lui attribuant la température 273·16 degrés Kelvin, exactement. (The 10CGPM decides to define the thermodynamic scale of temperature on which the triple point of water is a fixed fundamental point, attributing to it the temperature 273·16 K exactly.) This means that 1 K is 1/273·16 of the thermodynamic temperature of the triple point of water.
   Note: (1) The freezing point of pure air-free water ("standard temperature") is 273·15 K and its boiling point is 373·15 K (both under a pressure of one (standard) atmosphere).
   (2) The temperature T_K in K is related to the corresponding temperature θ_C in degrees Celsius by the formula:
   $T_K = \theta_C + 273·15$. $\theta_C = T_K - 273·15$.
   (3) The unit was formerly called the degree Kelvin. It has also been called the degree absolute (°A), but this refers to the temperature on any scale on which 0° corresponds to absolute zero (whatever the size of the intervals).
(2) *Quantity:* Temperature interval or difference (SI).
   Symbol: K; it was formerly deg and, prior to this, degK.
   Definition: 1 K is 1/273·16 of the interval between absolute zero and the triple point of water.
   Note: (4) 1 K is identical in size with the temperature interval one Celsius degree.
   (5) The original definition was: 1 K is one-hundredth of the interval between the freezing and boiling points of pure air-free water, both under a pressure of one (standard) atmosphere.
   (6) The unit was formerly called the Kelvin degree. It has also been called the absolute degree (degA), but this refers to the temperature interval of any scale on which 0° corresponds to absolute zero.
   (7) The former definition of the thermal volt made it equal to the kelvin.

KELVIN (3)

Quantity: Energy, particularly electrical energy (metric).
Definition: 1 kelvin = 1 kilowatt-hour.
Note: This was a name at one time applied to the unit now called the kilowatt-hour. It has also been called the Board of Trade unit.

KELVIN DEGREE degK

Quantity: Temperature interval or difference (SI).
Definition: 1 degK = 1 kelvin (2nd definition).
Note: (1) This is the former name of the unit now called the kelvin.
(2) See also the entry degree Kelvin.

KILO-...

See Appendix 3 section 2. The only units referred to in Part 1 are the following:

kilocalorie: see calorie (note 3)
kilogramme*
kilogramme-calorie*
kilogramme-equivalent*
kilogramme-force*

kilogrammetre*
kilogramme-weight*
kilohm: see ohm (note 3)
kiloton: see ton (1)
kilowatt-hour*

* Entered in its alphabetical position.

KILOGRAMME kg

Quantity: Mass (SI).
Definition: The official definition of this basic SI unit, given at the 3CGPM of 1901, is: Le kilogramme est l'unité de masse; il est représenté par la masse du prototype international du kilogramme. (The kilogramme is the unit of mass; it is represented by the mass of the international prototype kilogramme.)
Note: (1) The international kilogramme is a cylinder with a height equal to its diameter. It is made from an alloy of 90% platinum and 10% iridium, and is kept in the International Bureau of Weights and Measures at Sèvres, near Paris.
(2) The original definition of the kilogramme, recommended by the Paris Academy of Sciences in 1791, is: 1 kilogramme is the mass of one cubic decimetre (dm^3) of water at the temperature of its maximum density. 1 kg of water at this temperature (3·98 °C) was subsequently found to occupy a volume of 1·000 028 dm^3, and this volume was called the litre until its recent redefinition.
(3) The size of the kilogramme can be measured to an accuracy of more than 1 part in 10^8.

(4) The unit is popularly and loosely called the kilo. It was originally called the grave.

(5) In the US the name of this unit is spelled kilogram. The spelling in official translations of ISO Recommendations is always kilogramme.

KILOGRAMME-CALORIE   kcal (O)

Quantity: Heat energy (arbitrary).

Definition: One kilogramme-calorie is the quantity of heat required to raise the temperature of one kilogramme of air-free water from 14·5 °C to 15·5 °C at a constant pressure of one standard atmosphere. The experimentally derived value is 1 kcal = 4 185·5 ± 0·5 joules (J). 1 J ≈ 2·389 2 × 10^{-4} kcal.

Note: The unit is better regarded as one thousand fifteen-degree calories. It has also been called the large calorie.

KILOGRAMME-EQUIVALENT   kg-eq (O)

Quantity: Mass (metric-derived).

Definition: One kilogramme-equivalent is the mass of an element or radical in kilogrammes equivalent to (ie that combines with or replaces) three kilogrammes of tetravalent carbon-12.

KILOGRAMME-FORCE   kgf (BS)

Quantity: Force (MkgfS).

Definition: One kilogramme-force is the force which, when applied to a body of mass one kilogramme, gives it an acceleration equal to the standard acceleration of free fall (g_n). Since g_n = 9·806 65 metres per second squared, 1 kgf = 9·806 65 newtons (N). 1 N ≈ 0·101 972 kgf.

Note: In German-speaking countries this unit is called the Kilopond (unit symbol, kp).

KILOGRAMMETRE

Quantity: Energy (MkgfS).

Definition: One kilogrammetre is the work done when the point of application of a force of one kilogramme-force (kgf) is displaced through a distance of one metre (m) in the direction of the force.

1 kilogrammetre = 1 kgf m = 9·806 65 joules (J). 1 J ≈ 0·101 972 kilogram-metre.

KILOGRAMME-WEIGHT   kgwt (BS)

Quantity: Force (metric).

Definition: One kilogramme-weight is the force which, when applied to a body of mass one kilogramme (kg), gives it an acceleration equal to the local

value of the acceleration of free fall (g) expressed in metres per second squared.
1 kgwt = g newton (N). 1 N = $1/g$ kgwt.
Note: This is an inconsistent unit and its use is deprecated. The kilogramme-force should always be used in its place.

KILOWATT-HOUR   kWh

Quantity: Energy, particularly electrical energy (metric).
Definition: One kilowatt-hour is the energy expended when a power of one kilowatt (kW) is available for one hour (h). Since 1 kW = 1 000 W and 1 h = 3 600 seconds (s),
1 kWh = 1 kW h = 3.6×10^6 joules (J). 1 J $\approx 2.777\ 78 \times 10^{-7}$ kWh.
Note: The unit has also been called the Board of Trade unit and, at one time, the kelvin.

KINE

Quantity: Velocity (CGS).
Definition: One kine is the velocity at which a distance of one centimetre (cm) is traversed in one second (s). 1 kine = 1 cm/s = 0.01 m/s. 1 m/s = 100 kines.
Note: This unit has almost never been employed.

KINTAL

See quintal (2) (note)

KIP

Quantity: Mass (imperial).
Definition: 1 kip = 1 000 lb = 453.592 37 kg. 1 kg $\approx$ 0.002 204 62 kip.

KNOT

Quantity: Velocity, especially in aviation and navigation.
(1) Metric unit   kn (BS)
   Definition: One knot is the velocity at which a distance of one nautical mile (n mi) is traversed in one hour (h).
   1 kn = 1 n mi/h = 1 852/3 600 m/s $\approx$ 0.514 444 m/s. 1 m/s $\approx$ 1.943 84 kn.
   Note: (1) The unit is also called the international knot.
(2) Imperial unit   UKkn (BS)
   Definition: One UK knot is the velocity at which a distance of one UK nautical mile (UKn mi) is traversed in one hour.
   1 UKkn = 1 UKn mi/h = 6 080/3 600 ft/s $\approx$ 0.514 773 m/s.
   1 m/s $\approx$ 1.942 60 UKkn.
Note: (2) 1 kn $\approx$ 1.000 64 UKkn; 1 UKkn $\approx$ 0.999 361 kn.

LAMBDA λ (O)

Quantity: Volume (metric).
Definition: $1 \lambda = 1$ microlitre $= 10^{-6}$ litre $= 10^{-9}$ m^3 (present definition of the litre). 1 m$^3 = 10^9 \lambda$.

LAMBERT   L

Quantity: Luminance (metric).
Definition: One lambert is the luminance of a uniform diffuser emitting one lumen per square centimetre (cm^2). In terms of the candela (cd),

$$1 \text{ L} = \frac{1}{\pi} \text{ cd/cm}^2 = \frac{1}{\pi} \text{ stilb} = \frac{10^4}{\pi} \text{ nit (nt)}. \quad 1 \text{ nt} = 10^{-4}\pi \text{ L}.$$

Note: The use of this unit, although common in the US, is deprecated: the stilb should be used in its place.

LANGLEY

Quantity: Energy area-density, particularly of solar radiation (arbitrary).
Definition: One langley is the energy area-density of one calorie (cal$_{IT}$) of heat energy over one square centimetre (cm^2).
1 langley $= 1$ cal$_{IT}$/cm$^2 = 4 \cdot 186 \ 8 \times 10^4$ joules per metre squared (J/m^2).
1 J/m$^2 \approx 2 \cdot 388 \ 46 \times 10^{-5}$ langley.
Note: The unit was formerly defined as 1 langley $= 1$ cal$_{IT}$/cm^2 minute, but this was later called the pyron.

LEAGUE

Quantity: Length (imperial).
Definition: 1 league $= 3$ miles $= 4 \ 828 \cdot 032$ m. 1 m $\approx 2 \cdot 071 \ 24 \times 10^{-4}$ league.

LENTOR

Quantity: Kinematic viscosity (CGS).
Definition: One lentor is the kinematic viscosity of a fluid with a dynamic viscosity of one poise (P) and density one gramme per centimetre cubed (g/cm^3).
1 lentor $= 1$ P cm^3/g $= 1$ (g/cm s)cm^3/g $= 1$ cm^2/s $= 10^{-4}$ m^2/s.
1 m^2/s $= 10^4$ lentors.
Note: The lentor is a former name for the stokes, which should now always be used as the unit of kinematic viscosity (CGS).

LEO

Quantity: Acceleration (metric).
Definition: 1 leo $= 10$ metres per second squared (m/s^2). 1 m/s$^2 = 0 \cdot 1$ leo.
Note: This unit has rarely been employed.

LIGHT-WATT

Quantity: Luminous power. (This is generally regarded as an SI unit, although it belongs to all systems.)
Definition: One watt at a wavelength λ is equal to V_λ light-watts, where V_λ is the relative luminous efficiency of the radiation.
1 light-watt = 1 talbot per second = 1 lumberg per second.
Note: Values of V_λ are tabulated under the entry relativeluminous efficiency.

LIGHT YEAR

Quantity: Length (arbitrary).
Symbol: The unit symbol ly is often used.
Definition: One light year is the distance traversed in one year by electromagnetic waves in vacuo. The size of the light year depends on the type of year chosen and the experimentally determined sizes of the year and the velocity of electromagnetic waves in vacuo; a conveniently rounded value is
1 ly = $9{\cdot}460\ 5 \times 10^{15}$ m. 1 m $\approx 1{\cdot}057\ 0 \times 10^{-16}$ ly.
Note: This is a popular unit for astronomical measurements of distances of an order greater than those in the solar system.

LINE, METRIC

Quantity: Length (metric).
Definition: 1 metric line = 1 millimetre = 10^{-3} m. 1 m = 1 000 metric lines.
Note: See Appendix 4 for line.

LINE (OF ELECTRIC FORCE)

Quantity: Electric flux (CGS-esu)
Definition: One line (of electric force) is the electric flux associated with an electric charge of one franklin (Fr).
1 line (of electric force) = 1 Fr $\approx 3{\cdot}335\ 64 \times 10^{-10}$ coulomb (C).
1 C = $2{\cdot}997\ 925 \times 10^9$ lines (of electric force).

LINE (OF MAGNETIC FORCE)

Quantity: Magnetic flux (CGS-emu).
Definition: One line (of magnetic force) is the magnetic flux which, linking a circuit of one turn, produces in it an electromagnetic force of one abvolt (abV) as it is reduced to zero at a uniform rate in one second (s).
1 line (of magnetic force) = 1 abV s = 10^{-8} weber (Wb).
1 Wb = 10^8 lines (of magnetic force).
The CGS-emu mechanical equivalent is 1 line (of magnetic force) = 1 dyne½ cm.
The CGSm equivalent is 1 line (of magnetic force) = 1 erg/biot.

Note: The unit is better called the maxwell. It is alternatively called the abweber (abWb).

LINK li (O)

Quantity: Length (imperial).
Definition: 1 li = 0·01 chain = 7·92 inches = 0·201 168 m. 1 m ≈ 4·970 97 li.
Note: This unit is sometimes called Gunter's link or the surveyor's link.

LITRE l

Quantity: Capacity, ie volume (metric).
Definition: (1) The original definition, given at the 3CGPM of 1901, is: One litre is the volume occupied by a mass of one kilogramme of pure air-free water at the temperature of its maximim density and under a pressure of one (standard) atmosphere.
(2) At the 1950 CIPM the original definition was formalised to
1 l = 1·000 028 × 10^{-3} m³. 1 m³ ≈ 999·972 l.
(3) At the 12CGPM of 1964 it was resolved to recast the definition as:
1 l = 1 decimetre cubed = 0·001 m³. 1 m³ = 1 000 l.
Note: (1) It was agreed that the third definition should imply a special meaning of the word litre, and that it should not be used to express the results of accurate measurement. However, the third definition is now replacing the second one for scientific purposes. In the UK the Weights and Measures Act of 1963 still adheres to the second definition of the litre, although for practical purposes the third is likely to be used. In French law the third definition replaces the second one.
(2) The unit has also been called the kanne.
(3) In the US the name of this unit is spelled liter. The spelling in official translations of ISO Recommendations is always litre.

LITRE-ATMOSPHERE l atm

Quantity: Energy (metric).
Definition: One litre-atmosphere is the work done on a piston by a fluid under a pressure of one (standard) atmosphere (atm) when the volume swept out by the piston is one litre. Since 1 atm = 101 325 pascals, and using the present definition of the litre (10^{-3} m³), 1 l atm = 101 325 × 10^{-3} joule (J) = 101·325 J.
1 J ≈ 0·009 869 23 l atm.

LIVRE

Quantity: Mass (metric).
Definition: 1 livre = 0·5 kg. 1 kg = 2 livres.
Note: In French the word livre means pound. The Italian equivalent is libbra, the German equivalent is Pfund. 1 livre = 1·102 32 lb.

LOGIT

Quantity: Intensity level (all).
Definition: 1 logit = 1 decibel.
Note: This was one of several names proposed as alternatives for the decibel, but almost never employed.

LORENTZ UNIT

Quantity: Reciprocal length per magnetic flux density (arbitrary).
Definition: 1 Lorentz unit $= e/4\pi mc$, where e is the charge on the electron, m is the rest mass of the electron, and c is the velocity of electromagnetic radiation in vacuo. Hence 1 Lorentz unit $\approx$ 46·689 per tesla metre (/T m). 1 /T m $\approx$ 0·021 418 Lorentz units.
Note: The difference in position between a (zero-field) spectrum line and its Zeeman components, represented as a wave-number difference, is measured in Lorentz units. This unit is physically related to the Bohr magneton.

LOUDNESS UNIT LU (O)

Quantity: Loudness (all).
Definition: The loudness in loudness units is related to the loudness in sones and the equivalent loudness in phons such that for a sound of a certain frequency 1 LU $= 0·001$ sone $\equiv 0$ phon.
Note: This unit has fallen into disuse because the arbitrariness of the definition does not result in a formula connecting loudness in LU and in sones and equivalent loudness in phons.

LUMBERG

Quantity: Luminous energy. (This is generally regarded as a CGS unit, although it belongs to all systems.)
Definition: One lumberg is the luminous energy corresponding to one erg of radiant energy, having a luminous efficiency of one lumen (lm) for every erg per second (erg/s). 1 lumberg = 1 erg lm/(erg/s) = 1 lm s.
Note: This unit was formerly called the lumerg. It is equal in size to the talbot.

LUMEN lm

Quantity: Luminous flux (all).
Definition: One lumen is the flux emitted in a cone of solid angle one steradian (sr) by a spherical point source of uniform luminous intensity one candela (cd). 1 lm = 1 cd/sr.

LUSEC

Quantity: Leak rate, ie power (metric).
Definition: One lusec is a leak rate of one litre (l) per second (s) at a pressure of one millitorr. 1 lusec = 0·001 l torr/s. Using the new litre, 0·001 m³, and with 1 torr ≈ 133·322 pascals, 1 lusec ≈ 0·001 × 0·001 × 133·322 watt (W) ≈ 1·333 22 × 10⁻⁴ W. 1 W ≈ 7 500·62 lusecs.
Note: This unit is used for the measurement of the power of evacuation of a vacuum pump.

LUX lx

Quantity: (Intensity of) illumination (SI).
Definition: One lux is the illumination of one lumen (lm) uniformly over an area of one square metre (m²). 1 lx = 1 lm/m².
Note: The alternative name for this unit, the metre-candle, is deprecated.

LUXON

Quantity: Retinal illumination (metric).
Definition: One luxon is the retinal illumination produced by a surface having a luminance of one nit when the area of the pupil of the eye is one square millimetre.
Note: The unit is better called the troland. It was originally called the photon.

MACHE

Quantity: Radioactive concentration (arbitrary).
Definition: One mache is a concentration of 3·7 × 10⁻⁷ curie (Ci) of radio-active material in one cubic metre (m³) of a medium.
1 mache = 3·7 × 10⁻⁷ Ci/m³. 1 Ci/m³ ≈ 2·702 70 × 10⁶ maches.
Note: This unit is now obsolete.

MAGN

Quantity: Absolute permeability (SI).
Definition: 1 magn = 1 henry per metre.
Note: This unit has been proposed by the USSR, but has not received general acceptance.

MAGNETON

Quantity: Magnetic moment of an atomic particle (arbitrary).

(1) Bohr magneton β (BS)

Definition: 1 $\beta = eh/4\pi m$, where e is the charge on the electron, h is Planck's constant, and m is the rest mass of the electron. Hence:
1 $\beta \approx 9 \cdot 272\,9 \times 10^{-24}$ joule per tesla (J/T). 1 J/T $\approx 1 \cdot 078\,4 \times 10^{23}\,\beta$.

(2) Nuclear magneton

Definition: 1 nuclear magneton $= \dfrac{m}{M}\,\beta$, where M is the mass of the proton.

Since $M/m = 1\,836 \cdot 11$, 1 nuclear magneton $\approx 5 \cdot 050\,3 \times 10^{-27}$ J/T. 1 J/T $\approx 1 \cdot 980\,0 \times 10^{26}$ nuclear magnetons.

Note: The Bohr magneton has also been defined by 1 $\beta = \mu_0 eh/4\pi m$, where the magnetic constant $\mu_0 = 4\pi \times 10^{-7}$ henry per metre. Thus 1 $\beta \approx 1 \cdot 165\,3 \times 10^{-29}$ weber metre (Wb m). 1 Wb m $\approx 8 \cdot 581\,7 \times 10^{28}\,\beta$. It therefore follows that 1 nuclear magneton $\approx 6 \cdot 346\,6 \times 10^{-33}$ Wb m. 1 Wb m $\approx 1 \cdot 575\,7 \times 10^{32}$ nuclear magnetons.

MAGNITUDE   mag (O)

Quantity: Brightness of stars and other astronomical bodies (all).

Definition: The magnitude difference M of two stars relates the intensities I_1, I_2 of their light outputs (expressed in the same units) by the formula $M = 2 \cdot 5 \log (I_2/I_1)$. Thus 1 mag is the magnitude difference that corresponds to an intensity ratio of $10^{0 \cdot 4}$.

Note: The scale of magnitude is defined such that the brighter the star, the smaller its magnitude.

MAXWELL   Mx

Quantity: Magnetic flux (CGS-emu).

Definition: One maxwell is the magnetic flux which, linking a circuit of one turn, produces in it an electromotive force of one abvolt (abV) as it is reduced to zero at a uniform rate in one second (s). 1 Mx = 1 abV s = 10^{-8} weber (Wb). 1 Wb = 10^8 Mx. The CGS-emu mechanical equivalent is 1 Mx = 1 dyne$^{\frac{1}{2}}$ cm. The CGSm equivalent is 1 Mx = 1 erg/biot.

Note: An alternative name for the unit is the abweber (abWb). It is also called the line (of magnetic force).

MAYER

Quantity: Specific heat capacity (metric).

Definition: One mayer is equal to the quantity of heat in joules (J) required to raise the temperature of one gramme (g) of a substance by one kelvin (K). 1 mayer = 1 J/g K = 1 000 J/kg K. 1 J/kg K = 0·001 mayer.

McLEOD

Quantity: Pressure level. (The unit is derived from arbitrary units.)
Definition: The pressure level P in mcleods is related to the pressure p in millimetres of mercury (mmHg) by the formula $P = \log(1/p)$. Thus 1 mcleod is the pressure level that corresponds to a pressure of $0 \cdot 1$ mmHg.

MEGA-...

See Appendix 3 section 2. The only units referred to in Part 1 are the following:
    megaelectronvolt-curie (entered in its alphabetical position)
    megaton: see ton (1)
    megohm: see ohm (note 3)

MEGAELECTRONVOLT-CURIE   MeV Ci (O)

Quantity: Radioactive power (arbitrary).
Definition: One megaelectronvolt-curie is the power equivalent to that generated by one curie (Ci) emitting a mean energy of one megaelectronvolt (MeV) per disintegration.
1 MeV Ci = 10^6 eV × 1 Ci ÷ 1 disintegration ≈ $0 \cdot 005\ 927\ 77$ watt (W).
1 W ≈ $168 \cdot 698$ MeV Ci.

MEL

Quantity: Subjective pitch (all).
Definition: The pitch of a sound judged to be n times that of a tone of pitch 1 mel has a subjective pitch of n mels. At a sensation level of 40 decibels the pitch of a tone of frequency 1 000 hertz is 1 000 mels.
Note: The name of the unit is a shortened form of the word melody.

METRE   m

Quantity: Length (SI).
Definition: The official definition of this basic SI unit, given at the 11CGPM of 1960, is: Le mètre est la longueur égale à 1 650 763·73 longueurs d'onde dans le vide de la radiation correspondant à la transition entre les niveaux $2p_{10}$ et $5d_5$ de l'atome de krypton 86. (One metre is the length equal to 1 650 763·73 wavelengths in vacuo of the radiation corresponding to the transition between the energy levels $2p_{10}$ and $5d_5$ of the krypton-86 atom.) This corresponds to a wavelength of $6 \cdot 057\ 802\ 11 × 10^{-7}$ m.
Note: (1) The internationally-specified krypton discharge lamp is worked at a temperature of $63 \cdot 148$ K (the triple point of nitrogen).
(2) The original definition of the metre, recommended by the Paris Academy of Sciences in 1791, is: One metre is one ten-millionth of the length of the meridional quadrant of the earth's surface passing through Dunkirk and a

point close to Barcelona. This was replaced at the 7CGPM of 1927 by the then official definition: L'unité de longueur est le mètre, défini par la distance, à 0°, des axes des deux traits médians tracés sur la barre de platine iridié déposée au Bureau international des Poids et Mesures, et déclarée prototype du mètre par la Première Conférence Générale des Poids et Mesures, cette règle étant soumise à la pression atmosphérique normale et supportée par deux rouleaux d'au moins un centimètre de diamètre, situés symetriquement dans un même plan horizontal et à la distance de 571 mm l'un de l'autre. (The unit of length is the metre, defined by the distance at 0 °C of the axes of two median lines drawn on the platinum-iridium bar deposited at the International Bureau of Weights and Measures, and called the prototype metre by the 1CGPM, this measure being subject to standard atmospheric pressure and supported by two rollers of at least one centimetre in diameter, symmetrically placed in the same horizontal plane and at a distance of 571 mm from one another.) The prototype metre bar is 90% platinum and 10% iridium.

(3) A length of one metre is reproducible to 1 part in 10^8. Standardisation with a laser beam will probably increase the accuracy to 1 part in 10^{10}.

(4) The name stab has been proposed as an alternative for the metre, but never employed.

(5) In the US the name of this unit is spelled meter. The spelling in official translations of ISO Recommendations is always metre.

METRE-ATMOSPHERE   m-atm (O)

Quantity: Depth of equivalent atmosphere (arbitrary).

Definition: x m-atm of gas X is the depth (in metres) that an atmosphere would have if gas X were the only constituent and in the same amount as exists in the actual atmosphere, and reduced to standard temperature and pressure (s t p). The number of molecules in unit volume of a gas at s t p is given by the ratio of Avogadro's number ($6 \cdot 022\ 52 \times 10^{23}$ molecules/mole, international scale) to one Amagat volume unit: $2 \cdot 686\ 99 \times 10^{25}$ molecules/m^3. Thus, 1 m-atm $\equiv 2 \cdot 686\ 99 \times 10^{25}$ molecules/m^2. 1 molecule/m$^2 \approx 3 \cdot 721\ 63 \times 10^{-26}$ m-atm.

Note: The unit has also been called the atmo-metre.

METRE-CANDLE   mc (O)

Quantity: (Intensity of) illumination (SI).

Definition: One metre-candle is the illumination of one lumen (lm) uniformly over an area of one square metre (m^2). 1 mc = 1 lm/m^2.

Note: The name metre-candle is deprecated: the lux should be used in its place.

MHO ℧ (O)

Quantity: Conductance, admittance, susceptance (SI).
Definition: One mho is the conductance between two points of a conductor when a constant difference of potential of one volt (V) applied between these two points produces in this conductor a current of one ampère (A), the conductor not being the source of any electromotive force. 1 ℧ = 1 A/V.
Note: (1) The unit is better called the siemens; it has also been called the reciprocal ohm.
(2) This unit was formerly called the absolute mho (℧ abs), and must be distinguished from the international mho (℧int), formally abandoned in 1948: 1 ℧int = 1/1·000 49 ℧ ≈ 0·999 510 ℧. 1 ℧ = 1·000 49 ℧int.

MIC

Quantity: Inductance (metric).
Definition: 1 mic = 1 microhenry = 10^{-6} H. 1 H = 10^{6} mic.

MICRI-ERG

Quantity: Energy, particularly molecular surface energy (metric).
Definition: 1 micri-erg = 10^{-21} joule (J). 1 J = 10^{21} micri-ergs.
Note: This unit has rarely been employed.

MICRIL

Quantity: Concentration (metric).
Definition: One micril is a concentration of one milligramme (mg) of solute in one litre (l) of solvent. 1 micril = 1 mg/l. Using the new litre, 1 micril = 0·001 kilogramme per metre cubed (kg/m³). 1 kg/m³ = 1 000 micrils.
Note: This unit has also been called the gammil or the microgammil. It has rarely been used.

MICRO-...

See Appendix 3 section 2. The only units referred to in Part 1 are the following:

microbar: see barye (note 1)   microhm: see ohm (note 3)
microgammil: see gammil (note)   microlitre: see lambda
microhenry: see mic

MICROMETRE OF MERCURY (CONVENTIONAL)

See micron (2) (note 1)

MICRON (1)

Quantity: Length (metric).
Symbol: μm. The unit symbol was formerly μ.
Definition: 1 μm = 10^{-6} m. 1 m = 10^6 μm.
Note: This unit is now properly called the micrometre.

MICRON (2) μmHg

Quantity: Pressure (metric).
Definition: One micron is the pressure that would support a column of mercury of length one micrometre and density 13 595·1 kilogrammes per cubic metre under the standard acceleration of free fall (g_n). Since $g_n = 9\cdot806\,65$ m/s², 1 μmHg = 10^{-3} mmHg = 13 595·1 $\times$ 9·806 65 $\times$ 10^{-6} pascal (Pa) = 0·133 322 387 415 Pa. 1 Pa $\approx$ 7·500 62 μmHg.
Note: (1) This unit is properly called the (conventional) micrometre of mercury.
(2) The use of this unit is deprecated. It should be replaced by the millitorr, the size of which it equals to one part in seven million.

MIL (1) ml

Quantity: Capacity, ie volume (metric).
Definition: 1 ml = 10^{-3} litre. Using the present definition of the litre, 1 ml = 10^{-6} m³. 1 m³ = 10^6 ml.
Note: This unit, a contraction of the word millilitre, is used in pharmaceutical work.

MIL (2)

Quantity: Length (imperial).
Definition: 1 mil = 10^{-3} inch = 2·54 $\times$ 10^{-5} m. 1 m $\approx$ 3·937 01 $\times$ 10^4 mils.
Note: The unit is properly called the milli-inch. An alternative (popular) name is the thou.

MIL, ANGULAR

Quantity: (Geometrical) plane angle (all).
Definition: There are several alternative definitions.
(1) 1 angular mil = 0·001 radian (rad) = 9/50π ° $\approx$ 3′ 26·265″.
(2) 1 angular mil = 360/6 400 °   = π/3 200 rad = 3′ 22·5″.
(3) 1 angular mil = 90/1 000 °   = π/2 000 rad = 5′ 24″.

MIL, CIRCULAR

Quantity: Area (imperial).
Definition: One circular mil is the area of a circle of diameter one mil (10^{-3} in).

1 circular mil $= \dfrac{\pi}{4} \times 10^{-6}$ in$^2 \approx 7\cdot853\,98 \times 10^{-7}$ in$^2 \approx 5\cdot067\,07 \times 10^{-10}$ m^2.

1 m$^2 \approx 1\cdot973\,53 \times 10^9$ circular mils.

MILE

Quantity: Length (imperial).
Symbol: The unit symbol mi is useful and occasionally used, but it is not generally recognised.
Definition: 1 mi $=$ 1 760 yd $=$ 1 609·344 m. 1 m $\approx 6\cdot213\,71 \times 10^{-4}$ mi.
Note: In the UK the unit is legally termed the statute mile.

MILE, GEOGRAPHICAL

Quantity: Length (imperial).
Definition: 1 geographical mile $=$ 6 080 feet $\approx 1\cdot151\,52$ mile $=$ 1 853·184 m.
1 m $\approx 5\cdot396\,12 \times 10^{-4}$ geographical mile.
Note: This unit is now called the UK nautical mile.

MILE, NAUTICAL

Quantity: Length, especially in navigation.
(1) Metric unit   n mi (O)
    Definition: 1 n mi $=$ 1 852 m. 1 m $\approx 5\cdot399\,57 \times 10^{-4}$ n mi.
    Note: (1) The unit, also called the international nautical mile, was adopted by the International Hydrographic Conference of 1929, the UK and US dissenting. On 1 July 1954 the US adopted the definition, leaving the UK as the only dissenter.

(2) Imperial unit   UKn mi (O)
    Definition: 1 UKn mi $=$ 6 080 feet $\approx 1\cdot151\,52$ mile $=$ 1 853·184 m.
    1 m $\approx 5\cdot396\,12 \times 10^{-4}$ UKn mi.
    Note: (2) The original definition is: 1 UKn mi is the mean length of one minute of longitude. This was later formalised as:
    1 UKn mi $=$ 6 082 ft $\approx 1\cdot151\,89$ mile $=$ 1 853·793 6 m.
    The length 6 080 ft was called the geographical mile.
Note: (3) 1 n mi $\approx 0\cdot999\,363$ UKn mi. 1 UKn mi $\approx 1\cdot000\,64$ n mi.
(4) The unit must be distinguished from the sea mile.

MILE, SEA

Quantity: Length (imperial).
Definition: 1 sea mile = 1 000 fathoms = 6 000 feet = 1 828·8 m.
1 m ≈ 5·468 07 × 10⁻⁴ sea mile.
Note: This unit must be distinguished from the nautical mile.

MILE, TELEGRAPH NAUTICAL

Quantity: Length (imperial).
Definition: 1 telegraph nautical mile = 6 087 feet ≈ 1·152 84 mile = 1 855·317 6 m. 1 m ≈ 5·389 91 × 10⁻⁴ telegraph nautical mile.
Note: This unit is no longer in use.

MILLI-. . .

See Appendix 3 section 2. The only units referred to in Part 1 are the following:

milliampère-second*

milli-k: see k

millibar: see bar (note 1); vac

millilitre: see mil (1)

millicurie-destroyed*

millimass unit*

millicurie-hour*

millimetre of mercury (conventional)*

milliGal: see galileo

millimetre of water (conventional)*

milligramme-hour*

millitorr: see micron (2) (note)

milli-inch: see mil (2)

* Entered in its alphabetical position.

MILLIAMPÈRE-SECOND   mA s (O)

Quantity: Radiation dose due to exposure to X-rays (metric).
Definition: One milliampère-second is the product of the milliameter reading in milliampères (mA) and the exposure time in seconds (s).
1 mA s = 0·001 A × 1 s.
Note: The units milliampère-minute and millampère-hour are also sometimes employed.

MILLICURIE-DESTROYED   mcd (BS)

Quantity: Radiation dose due to exposure to X-rays (arbitrary).
Definition: One millicurie-destroyed is the amount of radiation emitted by a specimen of a radioactive nuclide during which its activity falls by one millicurie.
Note: For radon-222 (for which the mcd is most often employed) 1 mcd corresponds to 133 milligramme-hours approximately.

MILLICURIE-OF-INTENSITY-HOUR   Imch (O)

Quantity: Radiation dose due to exposure to γ-rays (arbitrary).
Definition: The experimentally determined size is 1 Imch = 8·38 röntgens approximately.
Note: The unit is better known as the sievert.

MILLIER

Quantity: Mass (MTS).
Definition: 1 millier = 1 000 kg. 1 kg = 0·001 millier.
Note: This unit is better called the tonne. It is also called the metric ton or the tonneau.

MILLIGRAMME-HOUR   mg h (O)

Quantity: Radiation dose due to exposure to γ-rays (arbitrary).
Definition: One milligramme-hour is the product of the equivalent radium content of the source in milligrammes (mg) and the exposure time in hours (h).
1 mg h = 0·001 g × 1 h.

MILLIHG

Quantity: Pressure (metric).
Definition: One millihg is the pressure that would support a column of mercury of length one millimetre and density 13 595·1 kilogrammes per cubic metre under the standard acceleration of free fall (g_n). Since g_n = 9·806 65 m/s², 1 millihg = 1 mmHg = 13 595·1 × 9·806 65 × 10^{-3} pascal (Pa) = 133·322 387 415 Pa. 1 Pa ≈ 0·007 500 62 millihg.
Note: The unit is better called the (conventional) millimetre of mercury. It is equal in size to the torr, to one part in seven million. The name is pronounced with the *h* silent.

MILLIMASS UNIT   mmu (BS)

Quantity: Mass (arbitrary).
Definition: One millimass unit is 1/16 000 of the atomic mass unit (physical scale). The experimentally derived value is 1 mmu = 1·037 38 × 10^{-31} kg.
1 kg ≈ 9·639 66 × 10^{30} mmu.

MILLIMETRE OF MERCURY (CONVENTIONAL)   mmHg

Quantity: Pressure (metric).
Definition: One (conventional) millimetre of mercury is the pressure that would support a column of mercury of length one millimetre and density 13 595·1 kilogrammes per cubic metre under the standard acceleration of

free fall (g_n). Since $g_n = 9.806\,65$ m/s², 1 mmHg = 13 595·1 × 9·806 65 × 10⁻³ pascal (Pa) = 133·322 387 415 Pa. 1 Pa ≈ 0·007 500 62 mmHg.
Note: The unit has also been called the millihg. It is equal in size to the torr, to one part in seven million.

MILLIMETRE OF WATER (CONVENTIONAL) mmH₂O

Quantity: Pressure (metric).
Definition: One (conventional) millimetre of water is the pressure that would support a column of water of length one millimetre and density 1 000 kilogrammes per cubic metre under the standard acceleration of free fall (g_n). Since $g_n = 9.806\,65$ m/s², 1 mmH₂O = 1 000 × 9·806 65 × 10⁻³ pascal (Pa) = 9·806 65 Pa. 1 Pa ≈ 0·101 972 mmH₂O.
Note: 1 mmH₂O = 1 kilogramme-force per metre squared.

MINIM

Quantity: Capacity, ie volume.
Symbol: min (BS). The sign ♏ was used in early times.
Definition: (1) UK unit (arbitrary)
   1 UKmin = $\frac{1}{480}$ UK fluid ounce ≈ 5·919 39 × 10⁻⁸ m³.
   1 m³ ≈ 1·689 36 × 10⁷ UKmin.
(2) US unit (imperial)
   1 USmin = $\frac{1}{480}$ US fluid ounce ≈ 6·161 15 × 10⁻⁸ m³.
   1 m³ ≈ 1·623 07 × 10⁷ USmin.
Note: The UK minim is used for the measurement of liquid and solid substances, although usually the former; the US minim is used only for the measurement of liquid substances.
1 UKmin ≈ 0·960 760 USmin; 1 USmin ≈ 1·040 84 UKmin.

MINUTE (of arc)

Quantity: (Geometrical) plane angle (all).
Definition: $1' = \frac{1}{60}°$ = $\pi/10\,800$ rad.
Note: (1) When using decimal fractions of a minute the unit symbol is placed after the figures, eg 7·5 '. In astronomical work it generally precedes the decimal point, eg 7 '·5.
(2) The unit has also been called the arcmin.

MINUTE (of time)

Quantity: Time (arbitrary).
Symbol: min. Where no confusion can arise (eg in ephimerides) m may be used. ' has also been used, but is not recommended.
Definition: 1 min = 60 seconds (s). 1 s ≈ 0·016 666 7 min.

MINUTE, CENTESIMAL ^{c} (O)

Quantity: (Geometrical) plane angle (none).
Definition: $1^c = 0.01$ grade $= \frac{1}{2}\pi \times 10^{-4}$ radian $= 0.009$ °.
Note: This unit has almost never been used. $1^c = 0.54'$.

MIRED

Quantity: Reciprocal colour temperature (metric).
Definition: One mired is the reciprocal colour temperature of a body which has a colour temperature of one million kelvins (K).
1 mired $= 10^6$ /K.   1 K $= 10^6$ /mired.
Note: The name of the unit is an abbreviation of the words micro-reciprocal degree.

MOHM

Quantity: Mechanical mobility (CGS).
Definition: One mohm is the ratio of a velocity of one centimetre per second (cm/s) to a force of one dyne (dyn).
1 mohm $= 1$ cm/dyn s $= 1\,000$ metres per newton second (m/N s).
1 m/N s $= 0.001$ mohm.
Note: The mohm is the reciprocal of the mechanical ohm. The name is an abbreviation of mobile ohm.

MOHR CUBIC CENTIMETRE   Mohr cm³ (O)

Quantity: Volume (arbitrary).
Definition: One Mohr cubic centimetre is the volume occupied by one gramme of pure air-free water at a specified temperature. Ie, 1 Mohr cm³ is numerically equal to the reciprocal of the density of water (at the specified temperature) expressed in grammes per centimetre cubed (g/cm³).
Note: This unit is used in saccharimetry. At the standardised temperature of 17.5 °C the experimentally derived value is 1 Mohr cm³ $\approx 1.000\ 13$ cm³ $\approx 1.000\ 13 \times 10^{-6}$ m³.   1 m³ $= 9.998\ 743\ 0 \times 10^5$ Mohr cm³.

MOLE   mol

Quantity: Amount of substance (see note 1).
Definition: One mole is the mass numerically equal (in grammes) to the relative molecular mass (molecular weight) of a substance. Ie, it is the amount of a substance that contains the same number of molecules as there are atoms in 0.012 kilogramme of carbon-12.
Note: (1) The unit will probably be accepted as the seventh basic SI unit.

(2) The unit was formerly called the mol or the gramme-molecule (mol or gmol).

(3) The mole is an individual unit of mass, ie it relates only to a given substance. If the relative molecular mass is μ, 1 mol $= \mu$ gramme (g), and the mass M of one mole is μ g/mol. Eg for hydrogen, $M_H = 2$ g/mol; for oxygen, $M_O = 32$ g/mol.

MORGAN

Quantity: Genetic map distance (arbitrary).
Definition: One morgan is the distance along the chromasome in a gene that gives a recombination frequency of one per cent.

MOUNCE

Quantity: Mass (metric).
Definition: 1 mounce $= 25$ g $= 0{\cdot}025$ kg. 1 kg $= 40$ mounces.
Note: The unit is also called the metric ounce; 1 mounce $\approx 0{\cdot}881\ 850$ ounce.

MUG

Quantity: Mass (metric-technical).
Definition: 1 mug $= 1$ metric slug.
Note: This name was proposed (like the par) as an alternative to the metric slug (metric-technical unit of mass) or hyl, but never employed.

MYRIA-. . .

See Appendix 3 section 2. Units are not entered under the prefix myria-.

NANO-. . .

See Appendix 3 section 2. Units are not entered under the prefix nano-.

NEPER

Quantity: Amplitude level, logarithmic decrement; in the special case explained in the note the unit measures intensity level (all).
Symbol: Np. The unit symbol N is generally used in telecommunications technology.
Definition: The amplitude level N in nepers relates two amplitudes a_1, a_2 (expressed in the same units) by the formula $N = \ln(a_2/a_1)$. Thus 1 Np is the amplitude level that corresponds to an amplitude ratio of e.
Note: Under those conditions in which the square of the amplitude of a vibration is proportional to the associated power there is a connexion between the Np and the decibel (dB): 1 Np $= 20 \log$ e dB $\approx 8{\cdot}685\ 89$ dB.

NEPIT

Quantity: Quantity of information (none)
Definition: The quantity of information I in nepits is related to the probability P at the receiver after the message is received and the probability P_o before reception by the formula $I = \ln (P/P_o)$.
Note: The unit is also called the nit. The name is a contraction of the term naperian digit.

NEUTRON RÖNTGEN

Quantity: Radiation dose due to neutrons (arbitrary).
Definition: One neutron röntgen is the dose of fast neutrons which, incident on a thimble ionisation chamber of specified characteristics, produces the same degree of ionisation as would one röntgen of γ- or X-rays.
Note: (1) This is an obsolete unit, also called the n-unit.
(2) A thimble ionisation chamber is a chamber in which the outer electrode is thimble-shaped.
(3) One neutron röntgen represents an absorbed dose in tissue of 2 to $2\frac{1}{2}$ rads approximately.

NEWTON   N

Quantity: Force (SI).
Definition: One newton is the force which, when applied to a body of mass one kilogramme (kg), gives it an acceleration of one metre per second squared (m/s²). $1 \text{ N} = 1 \text{ kg m/s}^2$.
Note: The unit was at one time called the large dyne.

NILE

Quantity: Reactivity (a measure of the departure of a nuclear reactor from its critical condition) (all).
Definition: One nile is the amount of reactivity equal to 0·01.
Note: This unit is rarely used.

NIT   nt   (1)

Quantity: Luminance (SI).
Definition: 1 nt = 1 candela per square metre (cd/m²).
Note: The use of this name is becoming rarer; the unit is often simply called the cd/m².

NIT (2)

Quantity: Quantity of information (none).
Definition: The quantity of information I in nits is related the probability P at the receiver after the message is received and the probability P_o before reception by the formula $I = \ln (P/P_o)$.
Note: The unit is also called the nepit. The name is a contraction of the term naperian digit.

NOX

Quantity: (Intensity of) illumination, particularly low-level illumination (metric).
Definition: 1 nox $= 10^{-3}$ lux (lx). 1 lx $=$ 1 000 nox.
Note: This unit was originally proposed as a scotoptic (dark-adapted eye) unit of illumination.

NOY

Quantity: Noisiness (all).
Definition: The noisiness N in noys is related to the perceived noise level L in perceived noise decibels (PNdB) by the formula $N = 2^{(L-40)/10}$. Thus 1 noy is the noisiness that corresponds to a perceived noise level of 40 PNdB.
Note: This unit is analogous to the sone.

n-UNIT

Quantity: Radiation dose due to neutrons (arbitrary).
Definition: One n-unit is the dose of fast neutrons which, incident on a thimble ionisation chamber of specified characteristics, produces the same degree of ionisation as would one röntgen of γ- or X-rays.
Note: (1) This is an obsolete unit, also called the neutron röntgen.
(2) A thimble ionisation chamber is a chamber in which the outer electrode is thimble-shaped.
(3) One n-unit represents an absorbed dose in tissue of 2 to $2\frac{1}{2}$ rads approximately.

OCTANT

Quantity: (Geometrical) plane angle (all).
Definition: 1 octant $= \pi/4$ rad $= 45°$.

OCTAVE

Quantity: (Pitch) interval (all).
Symbol: The unit symbol o is only used in submultiples.

Definition: The pitch interval I in octaves between two frequencies f_1, f_2 (expressed in the same units, and with f_2 greater than f_1) is given by the formula $I = \dfrac{1}{\log 2} \log (f_2/f_1) \approx 3{\cdot}321\ 93 \log (f_2/f_1)$.

OERSTED Oe

Quantity: Magnetic field strength (CGS-emu).
Definition: In a magnetic field of strength one oersted unit magnetic moment experiences a couple of one dyne centimetre (dyn cm). The unit of magnetic moment is therefore the dyn cm/Oe. 1 Oe $= 10^3/4\pi$ ampère/metre (A/m) $\approx$ 79·577 5 A/m. 1 A/m $\approx$ 0·012 566 4 Oe. The CGS-emu mechanical equivalent is 1 Oe $= 1$ dyn$^{\frac{1}{2}}$/cm. The CGSm equivalent is 1 Oe $= 1$ biot/cm.
Note: (1) The name oersted was proposed by the American Institute of Electrical Engineers in 1894 for the rationalised CGS-emu unit of reluctance. The IEC in 1930 gave it its present use.
(2) An alternative name for the unit is abampère per centimetre (abA/cm).

OHM Ω

Quantity: Resistance, reactance and impedance (SI).
Definition: One ohm is the resistance between two points of a conductor when a constant difference of potential of one volt (V) applied between these two points produces in this conductor a current of one ampère (A), the conductor not being the source of any electromotive force. $1\ \Omega = 1$ V/A.
Note: (1) Early formal definitions of the ohm were given in terms of the resistance of a column of mercury with the following properties:
(a) (Maxwell, 1868.) Length $= 1$ m; cross-sectional area $= 1$ mm^2. According to Maxwell, this makes $1\ \Omega$ equivalent to a mechanical unit of 10^7 m/s.
(b) (1st International Electrical Congress of 1881.) Temperature $= 0$ °C; length to be determined such that $1\ \Omega = 10^9$ CGS-emu units of resistance, a relationship defined by the British Association for the Advancement of Science in 1873.
(c) (1st International Electrical Congress, resumed in 1884.) Length $= 106$ cm. However, this did not quite make $1\ \Omega = 10^9$ CGS-emu units of resistance.
(d) (4th International Electrical Congress of 1893.) Temperature, that of melting ice; mass $= 14{\cdot}452\ 1$ g; length $= 160{\cdot}300$ cm; uniform cross section. To six significant figures, $1\ \Omega = 10^9$ CGS-emu units of resistance.
This last definition became law in the UK and the US in 1894; it was accepted in France (by decree) in 1896, and in Germany in 1898. It refers to a unit now called the international (Ω_{int}) to distinguish it from the ohm of the present-day formal definition, described as the absolute ohm (Ω_{abs}). By formal definition,

$1\,\Omega_{int} = 1{\cdot}000\,49\,\Omega; 1\,\Omega \approx 0{\cdot}999\,510\,\Omega_{int}$. An early name for the international ohm is the ohmad.

(2) The standard ohm has been experimentally determined with a Lorentz machine, a development of Faraday's wheel, to an accuracy (at the National Physical Laboratory) of 12 parts per million.

(3) The following multiples are irregularly named:

$$10^3 \ \ \Omega = 1 \text{ kilohm (not kilo-ohm)};$$
$$10^6 \ \ \Omega = 1 \text{ megohm (not megaohm)};$$
$$10^{-6}\,\Omega = 1 \text{ microhm (not micro-ohm)}.$$

OHMA

Quantity: Electric potential and potential difference, electromotive force (MKSA).
Definition: 1 ohma = 1 international volt.
Note: This unit is quite obsolete.

OHM, ACOUSTICAL Ω_a

Quantity: Acoustical resistance, reactance and impedance (CGS).
Definition: One acoustical ohm is the ratio of the sound pressure level of one dyne per square centimetre (dyn/cm²) to a source sound strength of one cubic centimetre per second (cm³/s).
$1\,\Omega_a = 1\ (dyn/cm^2)/(cm^3/s) = 1$ dyn s/cm⁵ $= 10^5$ pascals second per metre cubed (Pa s/m³). 1 Pa s/m³ $= 10^{-5}\,\Omega_a$.
Note: (1) The names ram and ray have been proposed as alternatives for the acoustical ohm, but never employed.
(2) The electrical ohm is an SI unit and not a CGS unit.

OHMAD

Quantity: Resistance, reactance and impedance (MKSA).
Definition: 1 ohmad = 1 international ohm.
Note: This unit is quite obsolete.

OHM, MECHANICAL Ω_m

Quantity: Mechanical resistance, reactance and impedance (CGS).
Definition: One mechanical ohm is the ratio of a force of one dyne (dyn) to a velocity of one centimetre per second (cm/s).
$1\,\Omega_m = 1$ dyn/(cm/s) $= 1$ dyn s/cm $= 0{\cdot}001$ newton second per metre (N s/m). 1 N s/m $= 1\,000\,\Omega_m$.
Note: (1) The names ram and ray have been proposed as alternatives for the mechanical ohm, but never employed.
(2) The mechanical ohm is the reciprocal of the mohm.
(3) The electrical ohm is an SI unit and not a CGS unit.

OHM, RECIPROCAL ℧

Quantity: Conductance, admittance, susceptance (SI).
Definition: One reciprocal ohm is the conductance between two points of a conductor when a constant different of potential of one volt (V) applied between these two points produces in this conductor a current of one ampère (A), the conductor not being the source of any electromotive force.
$1 \, ℧ = 1 \, A/V$.
Note: This name for the unit is deprecated; it should be called the siemens. An alternative name is the mho.

OHM, SPECIFIC ACOUSTICAL Ω_s

Quantity: Specific acoustical resistance, reactance and impedance (CGS).
Definition: One specific acoustical ohm is the ratio of a sound pressure level of one dyne per centimetre squared (dyn/cm^2) to a sound particle velocity of one centimetre per second (cm/s).
$1 \, \Omega_s = 1 \, (dyn/cm^2)/(cm/s) = 1 \, dyn \, s/cm^3 = 10$ pascals second per metre (Pa s/m). $1 \, Pa \, s/m = 0 \cdot 1 \, \Omega_s$.
Note: (1) This unit is also called the rayl. It was formerly called the unit-area acoustical ohm (Ω_u).
(2) The electrical ohm is an SI unit and not a CGS unit.

OHM, THERMAL

Quantity: Thermal resistance (SI).
Definition: One thermal ohm is the thermal resistance for which a temperature difference of one kelvin (K) causes an entropy flow of one watt (W) per kelvin. 1 thermal ohm $= 1 \, K^2/W$.
Note: (1) Formerly, one thermal ohm was the thermal resistance that allows a heat flow rate of one watt under a temperature difference of one kelvin. 1 thermal ohm $= 1 \, K/W$.
(2) The unit has also been called the fourier.

OPEN WINDOW UNIT owu (O)

Quantity: Equivalent absorption area of a surface to a sound (FPS).
Definition: One open window unit equals one square foot (ft^2) of a surface with a reverberation absorption coefficient of unity, which would absorb sound energy of a given frequency at the same rate as the surface under investigation. 1 owu $= 1 \, ft^2 = 0 \cdot 092 \, 903 \, 04 \, m^2$. $1 \, m^2 \approx 10 \cdot 763 \, 9$ owu.
Note: This unit is better called the sabin; it has also been called the (total) absorption unit.

OUNCE

Quantity: Mass (imperial).
(1) Avoirdupois measure   oz (BS)
   Definition: 1 oz = $\frac{1}{16}$ lb ≈ 0·028 349 5 kg. 1 kg ≈ 35·274 0 oz.
(2) Troy and apothecaries' measures
   Symbol: For troy ounce: oz tr in the UK, oz t in the US. For apothecaries'
   ounce: oz apoth in the UK, oz ap in the US; the sign ℥ was used in early
   times.
   Definition: 1 oz tr = 1 oz apoth = 480 grains = 0·031 103 476 8 kg.
   1 kg ≈ 32·150 7 oz tr = 32·150 7 oz apoth.
Note: See also the entry in Appendix 4.

OUNCEDAL

Quantity: Force (imperial).
Definition: One ouncedal is the force which, when applied to a body of mass
one ounce (oz), gives it an acceleration of one foot per second squared (ft/s²).
1 ouncedal = 1 oz ft/s² = 0·008 640 934 648 5 newton (N).
1 N ≈ 115·728 ouncedals.

OUNCE, FLUID   fl oz (BS)

Quantity: Capacity, ie volume.
Definition: (1) UK unit (arbitrary)
   1 UK fl oz = $\frac{1}{160}$ UK gallon = $\frac{1}{20}$ UK pint ≈ 2·841 30 × 10⁻⁵ m³.
   1 m³ ≈ 3·519 51 × 10⁴ UK fl oz.
(2) US unit (imperial)
   1 US fl oz = $\frac{1}{128}$ US gallon = $\frac{231}{128}$ in³ = $\frac{1}{16}$ US liquid pint =
   2·957 352 956 25 × 10⁻⁵ m³. 1 m³ ≈ 3·381 40 × 10⁴ US fl oz.
Note: The UK fluid ounce is used for the measurement of liquid and solid
substances, although usually the former; the US fluid ounce is used only for
the measurement of liquid substances.
1 UK fl oz ≈ 0·960 760 US fl oz; 1 US fl oz ≈ 1·040 84 UK fl oz.

OUNCE, METRIC

Quantity: Mass (metric).
Definition: 1 metric ounce = 25 g = 0·025 kg. 1 kg = 40 metric ounces.
Note: This unit is better known as the mounce. 1 metric ounce ≈ 0·881 850
ounce.

PAR

Quantity: Mass (metric-technical).
Definition: 1 par = 1 metric slug.

Note: This name was proposed (like the mug) as an alternative to the metric slug (metric-technical unit of mass) or hyl, but never employed.

PARKER

Quantity: Absorbed ionising radiation dose due to corpuscular radiation (ie α- and β-rays) (arbitrary).
Definition: One parker is the dose of ionising corpuscular radiation at which the energy absorbed by a substance is equal to the loss in energy during ionisation caused by one röntgen of electromagnetic radiation. One parker of corpuscular radiation is equivalent to one röntgen of electromagnetic radiation.
Note: The unit is better called the rep. It has also been called the tissue röntgen. The rad is used in preference to the parker.

PARSEC

Quantity: Length (arbitrary).
Symbol: The unit symbol pc is often used.
Definition: One parsec is the distance at which a base line of one astronomical unit (au) in length subtends an angle of 1 ″. 1 pc $= (6\cdot48 \times 10^5/\pi)$ au. Using the formally defined astronomical unit:
1 pc $\approx 3\cdot085\ 72 \times 10^{16}$ m. 1 m $\approx 3\cdot240\ 73 \times 10^{-17}$ pc.
Note: This unit is used for astronomical measurements of distances of an order greater than those in the solar system.

PASCAL   Pa

Quantity: Pressure (SI).
Definition: One pascal is the pressure resulting from a force of one newton (N) acting uniformly over an area of one square metre (m²). 1 Pa $= 1$ N/m².
Note: The unit has also been called the tor.

PASTILLE DOSE

Quantity: Radiation dose (arbitrary).
Definition: One pastille dose is the dose of radiation required to change the colour of a barium platinocyanide pastille from a specified apple-green colour (tint 'A') to a specified red-brown (tint 'B').
Note: This obsolete unit is also called the B-dose. It is equal to 500 röntgens approximately.

PECK   pk (O)

Quantity: Capacity, ie volume.
Definition: (1) UK unit (arbitrary)
    1 UKpk = 2 UK gallons ≈ 0·009 092 18 m³. 1 m³ ≈ 109·984 UKpk.
(2) US unit (imperial)
    1 USpk = 16 US dry pints = 0·008 809 767 541 72 m³.
    1 m³ ≈ 113·510 USpk.
Note: The UK peck is used for the measurement of solid and liquid substances, although usually the former; the US peck, before it became obsolete, was used only for the measurement of solid substances.
1 UKpk ≈ 1·032 06 USpk; 1 USpk ≈ 0·968 940 UKpk.

PENNYWEIGHT   dwt (BS)

Quantity: Mass (imperial).
Definition: 1 dwt = 24 grains = 1·555 173 84 × 10^{-3} kg.
1 kg ≈ 643·015 dwt.
Note: This is a unit of troy measure.

PERCH (1)

Quantity: Area (imperial).
Definition: 1 perch = 30¼ yd² = 25·292 852 64 m². 1 m² ≈ 0·039 536 9 perch.
Note: This unit, now obsolete, is more strictly called the square perch. It is also called the (square) rod and the (square) pole.

PERCH (2)

Quantity: Length (imperial).
Definition: 1 perch = 5½ yards = 5·029 2 m. 1 m ≈ 0·198 839 perch.
Note: This unit, now obsolete, is also called the rod or the pole.

PERMICRON

Quantity: Reciprocal length (metric).
Definition: One permicron is the reciprocal length of a distance which has a length of one micrometre (μm).
1 permicron = 1 /μm = 10^6 /m. 1 m = 10^6 /permicron.

PHON

Quantity: Equivalent loudness (all).
Definition: The equivalent loudness in phons is equal to the sound pressure level in decibels (above a datum level of 2 × 10^{-5} pascals root-mean-square), as judged by an otologically normal binaural listener, of a standard plane

sinusoidally-progressive pure tone of frequency 1 000 hertz coming from directly in front of the observer.

Note: The unit is related directly only to the sone.

PHOT

Quantity: (Intensity of) illumination (CGS).
Definition: One phot is the illumination of one lumen (lm) uniformly over an area of one square centimetre (cm²).
1 phot = 1 lm/cm² = 10^4 lux (lx). 1 lx = 10^{-4} phot.
Note: The alternative name for this unit, the centimetre-candle, is deprecated.

PHOTON (1)

Quantity: Energy associated with electromagnetic radiation (arbitrary).
Definition: The energy E in photons is related to the frequency f of the electromagnetic radiation in hertz by the formula $E = hf$, where h = Planck's constant in joule seconds (J s).
Note: (1) The experimentally derived size of h is $h = 6.624\,9 \times 10^{-34}$ J s.
(2) The unit is better called the quantum; it has also been called the ergon.

PHOTON (2)

Quantity: Retinal illumination (metric).
Definition: One photon is the retinal illumination produced by a surface having a luminance of one nit when the area of the pupil of the eye is one square millimetre.
Note: This was the original name for the unit now called the troland. It has also been called the luxon.

PICO-...

See Appendix 3 section 2. The only unit referred to in Part 1 is
     picofarad: see puff.

PIÈZE   pz (BS)

Quantity: Pressure (MTS).
Definition: One pièze is the pressure resulting from a force of one sthène (sn) acting uniformly over an area of one square metre (m²).
1 pz = 1 sn/m² = 1 000 pascals (Pa). 1 Pa = 0·001 pz.

PINT   pt (BS)

Quantity: Capacity, ie volume.
(1) UK pint (arbitrary)
     Definition: 1 UKpt = $\frac{1}{8}$ UK gallon ≈ 5·682 61 × 10^{-4} m³.
     1 m³ ≈ 1 759·76 UKpt.

Note: The UK pint, also called the imperial pint, is used for the measurement of liquid and solid substances, although usually the former.
1 UK pt $\approx$ 1·200 95 US liq pt $\approx$ 1·032 06 US dry pt.

(2) US liquid measure (imperial)

Definition: 1 US liq pt = $\frac{1}{8}$ US gallon = $28\frac{7}{8}$ in^3 = 4·731 764 73 × 10^{-4} m^3.
1 m^3 $\approx$ 2 113·38 US liq pt.

Note: The US liquid pint is used only for the measurement of liquid substances. 1 US liq pt $\approx$ 0·832 674 UK pt $\approx$ 0·859 367 US dry pt.

(3) US dry measure (imperial)

Definition: 1 US dry pt $= \frac{1}{64}$ US bushel $= \dfrac{107\ 521}{3\ 200}$ in$^3 \approx$ 5·506 10 × 10^{-4} m^3.

1 m$^3 \approx$ 1 816·17 US dry pt.

Note: The US dry pint is used only for the measurement of solid substances.
1 US dry pt $\approx$ 0·968 939 UK pt $\approx$ 1·163 65 US liq pt.

PLANCK

Quantity: Action (SI).
Definition: One planck is the action of energy of one joule (J) over one second (s). 1 planck = 1 J s.

PLI

Quantity: Line density (imperial).
Definition: One pli is the line density of a material which has a mass of one pound (lb) and a length of one inch (in), and is of uniform cross-section.
1 pli = 1 lb/in $\approx$ 17·858 0 kg/m. 1 kg/m $\approx$ 0·055 997 4 pli.

POINT

Quantity: Mass (metric).
Definition: 1 point = 0·01 metric carat = 2 × 10^{-6} kg. 1 kg = 5 × 10^5 points.
Note: This unit is only employed for commercial transactions in diamonds, pearls and other precious stones.

POISE P

Quantity: (Dynamic) viscosity (CGS).
Definition: One poise is the dynamic viscosity that gives rise to a tangential stress of one dyne per square centimetre (dyn/cm^2) across two planes separated by one centimetre when the velocity of streamlined flow is one centimetre per second (cm/s). 1 P = (1 dyn/cm^2) ÷ (cm/s)/cm = 1 dyn s/cm^2 = 0·1 newton second per metre squared (N s/m^2). 1 N s/m^2 = 10 P.

POISEUILLE   Pl (O)

Quantity: (Dynamic) viscosity (SI).
Definition: One poiseuille is the dynamic viscosity that gives rise to a tangential stress of one newton per metre squared (N/m²) across two planes separated by one metre when the velocity of streamlined flow is one metre per second (m/s). 1 Pl = 1 (N/m²) ÷ (m/s)/m = 1 N s/m².
Note: This unit has only rarely been used outside France.

POLE (1)

Quantity: Area (imperial).
Definition: 1 pole = $30\frac{1}{4}$ yd² = 25·292 852 64 m². 1 m² ≈ 0·039 536 9 pole.
Note: This unit, now obsolete, is more strictly called the square pole. It is also called the (square) rod and the (square) perch.

POLE (2)

Quantity: Length (imperial).
Definition: 1 pole = $5\frac{1}{2}$ yards = 5·029 2 m. 1 m ≈ 0·198 839 pole.
Note: (1) This unit, now obsolete, is also called the rod and the perch.
(2) In former times there existed various local poles with lengths of between 3 and 7 yards.

PONCELET

Quantity: Power (metric-derived).
Definition: One poncelet is the power available when a force of one hundred kilogrammes-force (kgf) is displaced through a distance of one metre (m) in the direction of the force during one second (s).
1 poncelet = 100 m kgf/s = 980·665 watts (W). 1 W ≈ 0·001 019 72 poncelet.

POUMAR

Quantity: Line density (imperial).
Definition: One poumar is the line density of a thread which has a mass of one pound (lb) and a length of one million yards (yd).
1 poumar = 10^{-6} lb/yd ≈ 4·960 55 × 10^{-7} kg/m. 1 kg/m ≈ 2·015 91 × 10^6 poumars.
Note: This unit is used in the textile industry as a measure of yarn count. The unit the tex is used in preference to the poumar.

POUND (1)   Lb

Quantity: Force (FSS, FlbfS).
Definition: 1 Lb = 1 pound-force ≈ 32·174 0 poundals = 4·448 221 615 260 5 newtons (N). 1 N ≈ 0·224 809 Lb.

Note: This unit, spelled Pound (with a capital letter), was a fundamental unit of the Stroud system (see Appendix 1 section 5–7). It is better called the pound-force.

POUND (2), (3)

Quantity: Mass.
(2) Avoirdupois measure (FPS) lb
  Definition: 1 lb = 0·453 592 37 kg. 1 kg ≈ 2·204 62 lb.
  Note: (1) The defining relation between the pound and the kilogramme has been chosen such that the ratio of the grain to the kilogramme is expressed as a terminating decimal. It was so defined in the US in the Federal Register of 1 July 1959, and thenceforth employed by the National Bureau of Standards and the American Standards Association in the US, and by the National Physical Laboratory and the British Standards Institution in the UK. It was legally adopted in the UK through the Weights and Measures Act of 1963.
  (2) The UK pound (UKlb) was originally defined in the Weights and Measures Act of 1878 as follows: The weight [sic] in vacuo of the platinum weight (as mentioned in the First Schedule to this Act), and by this Act declared to be the imperial standard for determining the imperial standard pound, shall be the legal standard measure of weight, and of measure having reference to weight, and shall be called the imperial standard pound, and shall be the only unit or standard measure of weight from which all other weights and all measures having reference to weight shall be ascertained. In 1933 the UKlb as defined by this Act was compared experimentally with the kilogramme: the result was
  1 UKlb = 0·453 592 338 kg.
  (?) The US pound (USlb avdp) was originally derived from the international kilogramme, and authorised in the Mendenhall Order of 5 April 1893 as 1 USlb advp = 0·453 529 427 7 kg.
(3) Troy and apothecaries' measures
  Symbol: For troy pound: lb tr or ℔ in the UK, lb t in the US. For apothecaries' pound: lb apoth in the UK, lb ap in the US.
  Definition: 1 lb tr = 1 lb apoth = 12 ounces (troy or apothecaries') = 5 760 grains = 0·373 241 721 6 kg. 1 kg ≈ 2·679 23 lb tr = 2·679 23 lb apoth.
  Note: The troy pound has legal standing in the US, but not in the UK.
Note: See Appendix 4 for great pound.

POUNDAL pdl

Quantity: Force (FPS).
Definition: One poundal is the force which, when applied to a body of mass

one pound (lb), gives it an acceleration of one foot per second squared (ft/s²).
1 pdl = 1 lb ft/s² = 0·138 254 954 376 newton (N). 1 N ≈ 7·233 01 pdl.

POUND-FORCE   lbf (BS)

Quantity: Force (FSS, FlbfS).
Definition: One pound-force is the force which, when applied to a body of
mass one pound, gives it an acceleration equal to the standard acceleration of
free fall (g_n). Alternatively, it may be defined as the force which, when applied
to a body of mass one slug, gives it an acceleration of one foot per second
squared (ft/s²).
Since g_n = 9·806 65 metres per second squared = 9·806 65/0·304 8 ft/s²
≈ 32·174 0 ft/s², 1 lbf ≈ 32·174 0 poundals = 4·448 221 615 260 5 newtons (N).
1 N ≈ 0·224 809 lbf.
Note: In the Stroud system this unit was called the Pound (spelled with a
capital letter).

POUND-WEIGHT   lbwt

Quantity: Force (imperial).
Definition: One pound-weight is the force which, when applied to a body of
mass one pound, gives it an acceleration equal to the local value of the
acceleration of free fall (g) expressed in feet per second squared.
1 lbwt = g poundal = 0·138 254 954 376 g newton (N). 1 N ≈ 7·233 01/g lbwt.
Note: This is an inconsistent unit and its use is deprecated. The pound-force
should always be used in its place.

PRA-. . .

The names of units of rationalised magnetic quantities in the SI international
system are obtained by prefixing the corresponding CGS-emu unit names by
pra- (representing the word 'practical'). The idea was suggested at the 1930
meeting of the IEC, but never taken up. The following table contains the four
units involved.

SI international unit	unit symbol	quantity measured	SI absolute unit now used
pragauss	praGs	magnetic flux density	tesla
pragilbert	praGb	magnetomotive force	ampère(-turn)
pramaxwell	praMx	magnetic flux	weber
praoersted	praOe	magnetic field strength	ampère per metre

PREECE

Quantity: Electrical resistivity (metric).
Definition: 1 preece = 10^{13} ohm metre (Ωm). 1Ωm = 10^{-13} preece.

PRISM DIOPTRE

Quantity: Deviating power of a prism (none).
Definition: The deviating power P of a prism in prism dioptres is related to the angle of deviation θ of a ray of light in any units by the formula $P = 100 \tan \theta$. Thus 1 prism dioptre is the deviating power that corresponds to an angle of deviation of 0·009 999 67 radian = 34' 23" approximately.

PROUT

Quantity: Nuclear binding energy (arbitrary).
Definition: One prout is the nuclear binding energy equal to one-twelfth of the binding energy of the deuteron. The experimentally derived value is 1 prout = 185·5 ± 0·1 kiloelectronvolts $\approx$ 2·971 4 × 10^{-14} joule (J). 1 J $\approx$ 3·365 5 × 10^{13} prouts.
Note: This unit has rarely been used.

PSI

Quantity: Pressure (imperial).
Definition: One psi is the pressure resulting from a force of one pound-force (lbf) acting uniformly over an area of one square inch (in^2).
1 psi = 1 lbf/in^2 $\approx$ 6 894·76 pascals (Pa). 1 Pa $\approx$ 1·450 38 × 10^{-4} psi.

PUFF

Quantity: Capacitance (metric).
Definition: 1 puff = 1 picofarad = 10^{-12} F. 1 F = 10^{12} puffs.

PYRON

Quantity: Power area-density (arbitrary).
Definition: One pyron is the power area-density that results from a thermal power of one calorie (cal_{IT}) per minute (min) acting uniformly over an area of one square centimetre (cm^2).
1 pyron = 1 cal_{IT}/cm^2 min = 697·8 watts per square metre (W/m^2).
1 W/m^2 $\approx$ 1·433 08 × 10^{-3} pyron.
Note: This unit was formerly called the langley.

QUADRANT (1)

Quantity: (Geometrical) plane angle (all).
Definition: 1 quadrant $= \pi/2$ rad $= 90°$.
Note: This unit is also called the right angle.

QUADRANT (2)

Quantity: Inductance (MKSA).
Definition: One quadrant is the inductance of a closed circuit which gives rise to a magnetic flux of one international weber (Wb_{int}) per international ampère (A_{int}). 1 quadrant $= 1$ Wb_{int}/A_{int}.
Note: (1) The 2nd International Electrical Congress of 1889 gave this name to the practical unit of coefficient of induction (a quantity now called inductance).
(2) The unit is better called the international henry. It has also been called the secohm.

QUADRANT (3)

Quantity: Length (metric).
Definition: 1 quadrant $= 10^7$ m. 1 m $= 10^{-7}$ quadrant.
Note: This unit was originally meant to be the length of the earth's meridional quadrant through Dunkirk and a point close to Barcelona. It is now obsolete.

QUANTUM

Quantity: Energy associated with electromagnetic radiation (arbitrary).
Definition: The energy E in quanta is related to the frequency f of the electromagnetic radiation in hertz by the formula $E = hf$, where h is Planck's constant in joule seconds (J s).
Note: (1) The experimentally derived size of h is $h = 6.624\ 9 \times 10^{-34}$ J s.
(2) The unit has also been called the photon and the ergon.

QUART

Quantity: Capacity, ie volume.
(1) UK unit (arbitrary)
  Definition: 1 UK quart $= 2$ UK pints $\approx 0.001\ 136\ 52$ m³.
  1 m³ ≈ 879.879 UK quarts.
  Note: The UK quart is used for the measurement of liquid and solid substances, although usually the former.
  1 UK quart $\approx 1.200\ 95$ US liquid quart $\approx 1.032\ 06$ US dry quart.
(2) US liquid measure (imperial)
  Definition: 1 US liquid quart $= 2$ US liquid pints $= 9.463\ 529\ 46 \times 10^{-3}$ m³.
  1 m³ $\approx 1\ 056.69$ US liquid quarts.

Note: The US liquid quart is used only for the measurement of liquid substances.

1 US liquid quart $\approx$ 0·832 674 UK quart $\approx$ 0·859 367 US dry quart.

(3) US dry measure (imperial)

Definition: 1 US dry quart = 2 US dry pints $\approx$ 0·001 101 22 m³.

1 m³ $\approx$ 908·085 US dry quarts.

Note: The US dry quart is used only for the measurement of solid substances.

1 US dry quart $\approx$ 0·968 939 UK quart $\approx$ 1·163 65 US liquid quart.

Note: (4) See Appendix 4 for reputed quart, winchester quart.

QUARTER (1)

Quantity: Capacity, ie volume (arbitrary).

Definition: 1 quarter = 8 UK bushels = 64 UK gallons $\approx$ 0·290 950 m³.

1 m³ $\approx$ 3·437 02 quarters.

Note: This is a UK unit for the measurement of liquid and solid substances; there is no corresponding US unit. See also the entry in Appendix 4.

QUARTER (2), (3), (4)

Quantity: Mass (imperial).

Definition: (2) UK avoirdupois measure   qr (BS)

1 qr = $\frac{1}{4}$ UK hundredweight = 28 lb = 12·700 586 36 kg.

1 kg $\approx$ 0·078 736 5 qr.

(3) US avoirdupois measure

1 quarter = $\frac{1}{4}$ US ton = 500 lb = 226·796 185 kg.

1 kg $\approx$ 0·004 409 25 quarter.

(4) Troy measure   qr tr (O)

1 qr tr = $\frac{1}{4}$ troy hundredweight = 25 troy pounds = 9·331 043 04 kg.

1 kg $\approx$ 0·107 169 qr tr.

QUINTAL   q (BS)

Quantity: Mass.

(1) Metric unit

Definition: 1 q = 100 kg. 1 kg = 0·01 q.

Note: This unit has also been called the metric centner.

(2) Imperial unit

Definition: 1 q = 100 lb = 45·359 237 kg. 1 kg $\approx$ 0·022 046 2 q.

Note: This unit, which has also been spelled kintal, has also been called the centner and the cental. In the US it is called the short hundredweight.

Q-UNIT   Q (Q)

Quantity: Potential heat energy of fuel reserves (arbitrary).
Definition: 1 Q = 10^{18} British thermal units $\approx 10^{21}$ joules (J). 1 J $\approx 10^{-21}$ Q.

RAD

Quantity: Absorbed ionising radiation dose (for inorganic matter) (arbitrary).
Definition: One rad is equivalent to an energy absorption per unit mass of
0·01 joule per kilogramme (J/kg) of irradiated material.
1 rad = 0·01 J/kg. 1 J/kg = 100 rads.
Note: For electromagnetic radiation in air, experiment has shown that
1 röntgen (R) is equivalent to 0·869 rad; 1 rad is equivalent to 1·15 R approximately. The numerical equivalent is different for other media.

RADIAN

Quantity: (Geometrical) plane angle (all).
Symbol: rad. The unit symbol ^{r} is often used in tables.
Definition: One radian is the angle which, having its vertex at the centre of a
circle, cuts off an arc equal in length to the radius.
1 rad = $180/\pi° \approx 57·295\ 8°$.

RADIATION LENGTH

Quantity: Length in cosmic-ray studies (arbitrary).
Definition: 1 radiation length = ln 2 shower units $\approx$ 0·693 147 shower
unit (qv).
Note: (1) This unit is also called the radiation unit or the cascade unit.
(2) The radiation length is an individual unit of length, ie its size depends on
the conditions of a given set of circumstances.

RAM

Quantity: Acoustical or mechanical resistance, reactance and impedance
(CGS).
Definition: 1 ram = 1 acoustical or mechanical ohm.
Note: This name (like the ray) was proposed as an alternative for either the
acoustical ohm or the mechanical ohm, but never employed.

RANKINE DEGREE   degR

Quantity: Temperature interval or difference (arbitrary).
Definition: 1 degR = 1 Fahrenheit degree = $\frac{5}{9}$ kelvin (K). 1 K = 1·8 degR.
Note: (1) The original definition was: 1 degR is $\frac{1}{180}$ of the interval between

the freezing and boiling points of pure air-free water, both under a pressure of one (standard) atmosphere.
(2) See also the entry degree Rankine.

RA-SIZE

Quantity: A size to which untrimmed paper is manufactured (metric).
Definition: (1) For reels of paper, the following are the standard sizes.

width	mm	430	610	860	1 220
imperial equivalent	in	16·93	24·02	33·86	48·03

(2) For sheets of paper, the following are the standard sizes.

designation	size mm × mm		imperial equivalent in × in	
RA0	860	1 220	33·86	48·03
RA1	610	860	24·02	33·86
RA2	430	610	16·93	24·02

Note: (1) The manufacturing tolerances allowed are ± 3 mm, ± 4 mm or ± 5 mm, whichever value is closer to 0·5%.
(2) Alternative metric series are the A-, B-, C- and SRA-sizes.
(3) When trimmed, RA-sizes correspond to A-sizes.

RAY

Quantity: Acoustical or mechanical resistance, reactance and impedance (CGS).
Definition: 1 ray = 1 acoustical or mechanical ohm.
Note: This name (like the ram) was proposed as an alternative for either the acoustical ohm or the mechanical ohm, but never employed.

RAYL

Quantity: Specific acoustical resistance, reactance and impedance (CGS).
Definition: One rayl is the ratio of a sound pressure of one dyne per square centimetre (dyn/cm^2) to a sound particle velocity of one centimetre per second (cm/s).
1 rayl = 1 $(dyn/cm^2)/(cm/s)$ = 1 dyn s/cm^3 = 10 pascals second per metre (Pa s/m). 1 Pa s/m = 0·1 rayl.
Note: The unit is also called the specific acoustical ohm. It was formerly called the unit-area acoustical ohm.

RAYLEIGH   R (O)

Quantity: Brightness, especially of the night sky and of the aurorae (arbitrary).

Definition: 1 R $= 10^{10}/4\pi$ quanta per (metre squared second steradian). Using the definition of the quantum in terms of the frequency f of the radiation in hertz, 1 R $\approx 5{\cdot}272\,0 \times 10^{-25}f$ watt per metre squared steradian (W/m² sr). 1 W/m² sr $\approx 1{\cdot}896\,8 \times 10^{24}/f$ R.

RÉAUMUR DEGREE   deg r (O)

Quantity: Temperature interval or difference (arbitrary).

Definition: 1 deg r $= 1{\cdot}25$ kelvin (K). 1 K $= 0{\cdot}8$ deg r.

Note: (1) The original definition was: 1 deg r is one-eightieth of the interval between the freezing and boiling points of pure air-free water, both under a pressure of one (standard) atmosphere.

(2) This unit is obsolete and has been abandoned.

(3) See also the entry degree Réaumur.

REM

Quantity: Absorbed ionising radiation dose (for organic matter), ie dose equivalent (arbitrary).

Definition: The dose in rems is numerically equal to the product of the dose in rads and certain modifying factors.

Note: (1) The only modifying factor known with any certainty and therefore not automatically put equal to unity is the quality factor. It possesses the following values:

    1   for β-rays with energies greater than 3×10^4 electronvolts, γ- and X-rays;

  1·7 for β-rays with energies less than 3×10^4 electronvolts;

  10   for α-rays and, usually, for neutrons of unknown energy;

  20   for atomic nuclei and fission fragments.

(2) The earlier definition, which virtually corresponds to the present one, is: 1 rem is the dose of radiation which, absorbed by living tissue, produces a biological effect equivalent to the action of one röntgen of electromagnetic radiation.

(3) The name is itself the unit symbol, since it stands for röntgen-equivalent mammal (or man).

(4) The unit has also been called the equivalent biological röntgen.

REP

Quantity: Absorbed ionising radiation dose due to corpuscular radiation (ie α- and β-rays) (arbitrary).

Definition: One rep is the dose of ionising corpuscular radiation at which the energy absorbed by a substance is equal to the loss in energy during ionisation caused by one röntgen of electromagnetic radiation. 1 rep of corpuscular radiation is equivalent to one röntgen of electromagnetic radiation.

Note: (1) The name is itself the unit symbol, since it standards for röntgen-equivalent physical.

(2) The unit has also been called the parker and the tissue röntgen.

(3) The rad is used in preference to the rep.

REYN

Quantity: (Dynamic) viscosity (FPS).

Definition: One reyn is the dynamic viscosity that gives rise to a tangential stress of one poundal per square foot (pdl/ft^2) across two planes separated by one foot when the velocity of streamlined flow is one foot per second (ft/s). 1 reyn = 1 (pdl/ft^2) ÷ (ft/s)/ft = 1 pdl s/ft^2 = 1·488 16 newton second per metre squared (N s/m^2). 1 N s/m^2 ≈ 0·671 969 reyn.

RHE

Quantity: Fluidity, ie reciprocal viscosity (metric).

Definition: (1) Dynamic fluidity

One rhe is the dynamic fluidity of a fluid which has a dynamic viscosity of one centipose (cP).

1 rhe = 1 /cP = 1 000 /(N s/m^2) = 1 000 m^2/N s.

1 N s/m^2 (newton second per metre squared) = 1 000 /rhe.

(2) Kinematic fluidity

One rhe is the kinematic fluidity of a fluid which has a kinematic viscosity of one centistokes (cSt).

1 rhe = 1 /cSt = 10^6 s/m^2. 1 m^2/s = 10^{-6} /rhe.

RICHTER MAGNITUDE *M* (O)

Quantity: Intensity of earthquakes. (The unit measures a dimensionless quantity and thus belongs to no system.)

Definition: The magnitude of an earthquake *M* on the Richter scale is given by the formula $\log E = a + bM$, where E = total energy released and a and b are constants, the values of which are modified through observations. With E in joules, good working values of the constants are $a = -1·2$, $b = +2·4$.

Note: The following table relates values of E to M from the formula and constants given above.

M	E (in joules)
0	$6\cdot3 \times 10^{-2}$
1	$1\cdot6 \times 10$
2	$4\cdot0 \times 10^3$
3	$1\cdot0 \times 10^6$
4	$2\cdot5 \times 10^8$
5	$6\cdot3 \times 10^{10}$
6	$1\cdot6 \times 10^{13}$
7	$4\cdot0 \times 10^{15}$
8	$1\cdot0 \times 10^{18}$
9	$2\cdot5 \times 10^{20}$
10	$6\cdot3 \times 10^{22}$

RIGHT ANGLE

Quantity: (Geometrical) plane angle (all).
Symbol: $\llcorner$. Rt $\llcorner$ (BS) has also been used.
Definition: $1 \llcorner = \pi/2$ rad $= 90°$.
Note: The unit is also called the quadrant.

ROD (1)

Quantity: Area (imperial).
Definition: 1 rod $= 30\frac{1}{4}$ yd² $= 25\cdot292\ 852\ 64$ m². 1 m² $\approx 0\cdot039\ 536\ 9$ rod.
Note: This unit, now obsolete, is more strictly called the square rod. It is also called the (square) pole and the (square) perch.

ROD (2)

Quantity: Length (imperial).
Definition: 1 rod $= 5\frac{1}{2}$ yards $= 5\cdot029\ 2$ m. 1 m $\approx 0\cdot198\ 839$ rod.
Note: This unit, now obsolete, is also called the pole and the perch.

RÖNTGEN

Quantity: Radioactive dose due to exposure to electromagnetic radiation (ie γ- and X-rays) (arbitrary).
Symbol: R (BS); formerly r (BS).
Definition: One röntgen is the dose of electromagnetic radiation which will produce in air a charge of $2\cdot58 \times 10^{-4}$ coulombs (C) on all the ions of one sign, when all the electrons of both signs liberated in a volume of air of mass one kilogramme (kg) are stopped completely.
1 R $= 2\cdot58 \times 10^{-4}$ C/kg. 1 C/kg $\approx 3\ 880$ R.
Note: (1) This definition makes the unit identical with that given by the former definition: the dose of electromagnetic radiation which will produce

ions carrying a charge of one franklin (of each sign) per 0·001 293 grammes of dry air at standard temperature and pressure (stp), ie per cubic centimetre of dry air.

(2) The original definition was: the dose of electromagnetic radiation which, when the secondary electrons are fully utilised and the wall effect of the chamber is avoided, produces in one cubic centimetre of atmospheric air at a temperature of 0 °C and 76 centimetres of mercury pressure such a degree of conductivity that one franklin of charge is measured at saturation current.

(3) A dose of 1 R corresponds to $2·082 \times 10^{15}$ ion-pairs per cubic metre of dry air at stp. It also corresponds to 0·008 69 joules absorbed per kilogramme of air, and to slightly less than 0·01 joules absorbed per kilogramme of water or living matter. (All the values quoted here have been experimentally determined.)

(4) The unit name is also spelled roentgen.

RÖNTGEN, EQUIVALENT BIOLOGICAL EBR (O)

Quantity: Absorbed ionising radiation dose (for organic matter), ie dose equivalent (arbitrary).

Definition: One equivalent biological röntgen is the dose of radiation which, absorbed by living tissue, produces a biological effect equivalent to the action of one röntgen of electromagnetic radiation.

Note: The unit is now called the rem, and the definition has been modified (see rem).

RÖNTGEN-EQUIVALENT MAMMAL (or MAN)

See rem

RÖNTGEN-EQUIVALENT PHYSICAL

See rep

RÖNTGEN-PER-HOUR-AT-ONE-METRE rhm (BS)

Quantity: Radiation dose rate (ie intensity) due to γ-rays (arbitrary).

Definition: One röntgen-per-hour-at-one-metre is the intensity of a γ-ray source (under specified conditions of shielding) such that, at a distance of one metre in air, its γ-rays produce a dose rate of one röntgen per hour.

Note: 1 rhm = 1 curie approximately.

RÖNTGEN, TISSUE

Quantity: Absorbed ionising radiation dose due to corpuscular radiation (ie α- and β-rays) (arbitrary).

Definition: One tissue röntgen is the dose of ionising corpuscular radiation at which the energy absorbed by a substance is equal to the loss in energy during ionisation caused by one röntgen of electromagnetic radiation. One tissue röntgen of corpuscular radiation is equivalent to one röntgen of electromagnetic radiation.

Note: The unit is better called the rep. It has also been called the parker. The rad is used in preference to the tissue röntgen.

ROOD

Quantity: Area (imperial).
Definition: 1 rood = $\frac{1}{4}$ acre = 1 210 yd² = 1 011·714 105 6 m².
1 m² ≈ 9·884 22 × 10^{-4} rood.

ROWLAND

Quantity: Wavelength (arbitrary).
Definition: The experimentally derived size is 1 rowland = 999·81/999·94 ångström (Å) ≈ 0·999 87 Å ≈ 9·998 7 × 10^{-11} m. 1 m ≈ 1·000 1 × 10^{10} rowland.

Note: The unit has long been obsolete, having been replaced by the ångström; the nanometre is now preferred. Its use arose from an error in the original measurements of wavelength using the ångström.

RUM

Quantity: Pressure (CGS).
Definition: 1 rum = 1 barye = 1 dyn/cm² = 1 microbar.
Note: This name was proposed as an alternative for the barye (dyn/cm²), but never employed. It was formerly called the barad.

R-UNIT

Quantity: Radiation dose rate (ie intensity) due to X-rays (arbitrary).
Definition: (1) Solomon R-unit
   One Solomon R-unit is the intensity of X-radiation from a source equal to that from one gramme of radium placed two centimetres from an ionisation chamber, there being a platinum screen one-half a millimetre thick between the radium and the chamber. The unit is now standardised as 1 Solomon R-unit = 2 100 röntgens per hour ≈ 0·583 333 röntgen per second (R/s). 1 R/s ≈ 1·714 29 Solomon R-unit.
(2) German R-unit
   1 German R-unit = 2·5 Solomon R-units approximately.
   1 German R-unit ≈ 1$\frac{1}{2}$ R/s. 1 R/s ≈ 0·7 German R-unit.

RUTHERFORD   rd (BS)

Quantity: Radioactive disintegration rate, ie activity (metric).
Definition: One rutherford is the quantity of a radioactive nuclide required
to produce one million disintegrating atoms per second, or, simply, one
million disintegrations per second.
1 rd $\approx$ 2·702 70 $\times$ 10^{-5} curie (Ci). 1 Ci = 3·7 $\times$ 10^4 rd.
Note: The curie is used in preference to the rutherford.

RYDBERG (1)

Quantity: Energy, particularly in atomic studies (Hartree system of units).
Definition: 1 rydberg = $e^2/2a_0$, where e is the atomic unit of charge and a_0
is the atomic unit of length. Hence 1 rydberg $\approx$ 2·425 2 $\times$ 10^{-18} joule (J).
1 J $\approx$ 4·123 3 $\times$ 10^{17} rydbergs.
Note: This unit is also called the atomic unit of energy.

RYDBERG (2)

Quantity: Reciprocal length, especially wave number (CGS).
Definition: One rydberg is the reciprocal length of a distance which has a
length of one centimetre (cm).
1 rydberg = 1 /cm = 100 /m. 1 m = 100 /rydberg.
Note: This unit is better called the kayser. It has also been called the balmer.

SABIN

Quantity: Equivalent absorption area of a surface to a sound (FPS).
Definition: One sabin equals one square foot (ft^2) of a surface with a reverberation absorption coefficient of unity, which would absorb sound energy at
the same rate as the surface under investigation.
1 sabin = 1 ft^2 = 0·092 903 04 m^2. 1 m^2 $\approx$ 10·763 9 sabins.
Note: This unit has also been called the (total) absorption unit or the open
window unit.

SAVART   s (O)

Quantity: (Pitch) interval (all).
Definition: The pitch interval I_s in savarts between two frequencies f_1, f_2
(expressed in the same units, and with f_2 greater than f_1) is given by the
formula I_s = 1 000 log (f_2/f_1).
Note: I_s is related to the corresponding pitch interval I_o in octaves by
I_s = 1 000 log 2 I_o $\approx$ 301·030 I_o; I_o $\approx$ 0·003 321 93 I_s.

SAVART, MODIFIED

Quantity: (Pitch) interval (all).
Definition: The pitch interval I_{ms} in modified savarts between two frequencies f_1, f_2 (expressed in the same units, and with f_2 greater than f_1) is given by the formula $\quad I_{ms} = \dfrac{300}{\log 2} \log (f_2/f_1) \approx 996 \cdot 594 \log (f_2/f_1).$

Note: I_{ms} is related to the corresponding pitch intervals I_o in octaves and I_s in savarts by $I_{ms} = 300\ I_o \approx 0 \cdot 996\ 594\ I_s$; $I_o \approx 0 \cdot 003\ 333\ 33\ I_{ms}$; $I_s \approx 1 \cdot 003\ 43\ I_{ms}$.

SCRUPLE

Quantity: Mass (imperial).
Symbol: There is no unit symbol, although the sign $\ni$ was used in early times.
Definition: 1 scruple = 20 grains = $1 \cdot 295\ 978\ 2$ kg. 1 kg $\approx 7\ 716 \cdot 18$ scruples.
Note: This is a unit of apothecaries' measure.

SECOHM

Quantity: Inductance (MKSA).
Definition: One secohm is the inductance of a closed circuit which gives rise to a magnetic flux of one international weber (Wb_{int}) per international ampère (A_{int}). 1 secohm = 1 Wb_{int}/A_{int}.
Note: This unit is better called the international henry. It has also been called the quadrant.

SECOND (of arc)   ″

Quantity: (Geometrical) plane angle (all).
Definition: $1'' = \frac{1}{360}^{\circ} = \pi/648\ 000$ rad.
Note: When using decimal fractions of a second the unit symbol is placed after the figures, eg $7 \cdot 5$ ″. In astronomical work it generally precedes the decimal point, eg 7 ″·5.

SECOND (of time)

Quantity: Time (all).
Symbol: s.   ″ has also been used, but is not recommended.
Definition: The official definition of this basic SI unit, given at the 12CGPM of 1964, is: 1 second is the interval occupied by 9 192 631 770 cycles of the radiation corresponding to the $(F = 4, M_F = 0) - (F = 3, M_F = 0)$ transition of the caesium-133 atom when unperturbed by exterior fields.
Note: (1) The original definition of the second was: 1 second is the fraction 1/86 400 of a mean solar day. This is now called the mean solar second. One

mean solar day is the time interval between consecutive passages of the sun across the meridian, averaged over one year. This definition was replaced at the 11CGPM of 1960 by: La seconde est la fraction 1/31 556 925·974 7 de l'année tropique pour 1900 janvier 0 à 12 heures de temps des éphémérides. (The second is the fraction 1/31 556 925·974 7 of the tropical year for 1900 January 0 at 12 hours ephemeris time.) One tropical year is the time interval between consecutive passages, in the same direction, of the sun through the earth's equatorial plane. This latter definition now defines the ephemeris second.

(2) The standard second is experimentally reproducible using an Essen-ring quartz clock to an accuracy of 1 part in 10^{12}.

(3) The sidereal second is defined as the fraction 1/86 400 of the time for one complete rotation of the earth on its axis. The experimentally observed size is 1 sidereal second = 86 164·098 92/86 400 s $\approx$ 0·997 270 s.   1 s $\approx$ 1·002 74 sidereal second.

(4) See also the entry in Appendix 4.

SECOND, CENTESIMAL ^{cc} (O)

Quantity: (Geometrical) plane angle (none).
Definition: 1 ^{cc} = 10^{-4} grade = $\frac{1}{2} \pi \times 10^{-6}$ radian = 9×10^{-5} °.
Note: This unit has almost never been used. 1 ^{cc} = 0·324 ".

SEXTANT

Quantity: (Geometrical) plane angle (all).
Definition: 1 sextant = $\pi/3$ rad = 60 °.

SHED

Quantity: Area, especially the cross-sectional area of an atomic nucleus (metric).
Definition: 1 shed = 10^{-24} barn = 10^{-48} cm² = 10^{-52} m².
1 m² = 10^{52} sheds.
Note: This unit has almost never been used.

SHOWER UNIT

Quantity: Length in cosmic-ray studies (arbitrary).
Definition: One shower unit is the mean path length required to reduce the energy of a charged particle by one-half.
Note: The shower unit is an individual unit of length, ie its size depends on the conditions of a given set of circumstances.

SIEGBAHN UNIT

Quantity: Length, particularly wavelength in X-ray spectra (arbitrary).
Symbol: X. The unit symbol XU has also been used.
Definition: The experimentally derived value was subsequently formalised by
assigning the value 3 029·45 X to the spacing of the (200) planes of calcite
at 18 °C. This gives 1 X $= (1·002\ 02 \pm 0·000\ 03) \times 10^{-13}$ m.
1 m $\approx 9·979\ 84 \times 10^{12}$ X.
Note: The unit is also called the X-unit or the X-ray unit. It was abandoned
in 1948 in favour of the ångström, which was itself subsequently abandoned
in favour of the nanometre.

SIEMENS S

Quantity: Conductance, admittance, susceptance (SI).
Definition: One siemens is the conductance between two points of a conductor when a constant difference of potential of one volt (V) applied between
these two points produces in this conductor a current of one ampère (A), the
conductor not being the source of any electromotive force. 1 S $= 1$ A/V.
Note: (1) The unit is also called the mho (reciprocal ohm).
(2) This unit was formerly called the absolute siemens (S_{abs}), and must be
distinguished from the international siemens (S_{int}), formally abandoned in
1948: 1 $S_{int} = 1/1·000\ 49$ S $\approx 0·999\ 510$ S. 1 S $= 1·000\ 49\ S_{int}$.
(3) The name siemens was used in Germany up to 1930 for the unit of
resistance; its size was approximately equal to that of the international ohm.

SIEVERT

Quantity: Radiation dose due to γ-rays (arbitrary).
Definition: One sievert is the dose of γ-rays delivered in one hour at a dis-
tance of one centimetre from a point source of one milligramme of radium
enclosed in a platinum container with walls of thickness half a millimetre.
The experimentally determined size is 1 sievert $= 8·38$ röntgens approximately.
Note: The unit was originally called the millicurie-of-intensity-hour.

SIGN

Quantity: (Geometrical) plane angle (all).
Definition: 1 sign $= \pi/6$ radian $= 30$ °.
Note: This unit represents the mean angular extent of one sign of the zodiac.

SIRIOMETER

Quantity: Length (arbitrary).
Definition: (1) 1 siriometer $= 10^6$ astronomical units (au). Using the formally
defined au, 1 siriometer $= 1·496\ 00 \times 10^{17}$ m. 1 m $\approx 6·684\ 49 \times 10^{-18}$
siriometer.

(2) 1 siriometer = 5 parsecs $\approx$ 1·542 86 × 10¹⁷ m. 1 m $\approx$ 6·481 47 × 10⁻¹⁸ siriometer.

Note: This obsolete unit was used for astronomical measurements of distance of an order greater than those in the solar system. When first proposed, the siriometer was meant to represent the distance of the star Sirius. In fact, Sirius is now known to be at a distance of 2·7 parsecs.

SKIN ERYTHEMA DOSE   SED (O)

Quantity: Radioactive dose due to exposure to electromagnetic radiation (ie γ- and X-rays) (arbitrary).

Definition: One skin erythema dose is the dose of electromagnetic radiation which, in eighty per cent of the cases, slightly reddens or browns the skin within the three weeks after exposure. The experimentally determined size is:

1 skin erythema dose = 1 000 röntgens approximately (for γ-rays);
                                            600 röntgens approximately (for X-rays).

SKOT

Quantity: Luminance, particularly low-level luminance (metric).

Definition: 1 skot = 10⁻³ apostilb = 10⁻³ /π candela per metre squared (ie nit, nt). 1 nt = 1 000π skot.

Note: The use of this unit is deprecated; the millinit should be used in its place. It was originally proposed as a scotoptic (dark-adapted eye) unit of luminance.

SLUG

Quantity: Mass (FSS).

Definition: One slug is the mass that acquires an acceleration of one foot per second squared under the influence of a force of one pound-force.

1 slug = 9·806 65/0·304 8 lb $\approx$ 32·174 0 lb $\approx$ 14·593 9 kg.

1 kg $\approx$ 0·068 521 8 slug.

Note: This is the British technical unit of mass. It has also been called the gee pound.

SLUG, METRIC

Quantity: Mass (metric-technical).

Definition: One metric slug is the mass that acquires an acceleration of one metre per second squared under the influence of a force of one kilogramme-force. 1 metric slug = 9·806 65 kg. 1 kg $\approx$ 0·101 972 metric slug.

Note: The name for this unit is not often used; it is generally simply referred to as the metric-technical unit of mass. It is also called the hyl. Alternative proposed (but unemployed) names are the mug and the par.

SONE

Quantity: Loudness (all).
Definition: The loudness S in sones is related to the equivalent loudness P in phons by the formula $S = 2^{(P-40)/10}$. Thus 1 sone is the loudness that corresponds to an equivalent loudness of 40 phons.
Note: (1) The unit is defined so as to give scale values approximately proportional to the observed loudness of the sound. The formula has been justified by experiment over the range 20 phons to 120 phons.
(2) The unit is analogous to the noy.

SPAT S (O)

Quantity: Length (metric).
Definition: $1 \text{ S} = 10^{12}$ m. $1 \text{ m} = 10^{-12}$ S.
Note: This rarely-used unit has been employed for astronomical measurements of distance, normally confined to the solar system.
$1 \text{ S} \approx 6 \cdot 684\,49$ astronomical units.

SRA-SIZE

Quantity: A size to which untrimmed paper is manufactured (metric).
Definition: (1) For reels of paper, the following are the standard sizes.

width	mm	450	640	900	1 280
imperial equivalent	in	17·72	25·20	35·43	50·39

(2) For sheets of paper, the following are the standard sizes.

designation	size mm × mm		imperial equivalent in × in	
SRA0	900	1 280	35·43	50·39
SRA1	640	900	25·20	35·43
SRA2	450	640	17·72	25·20

Note: (1) The manufacturing tolerances allowed are ± 3 mm, ± 4 mm or ± 5 mm, whichever value is closer to $0 \cdot 5 \%$.
(2) Alternative metric series are the A-, B-, C- and RA-sizes.
(3) When trimmed, SRA-sizes correspond to A-sizes.

STAB

Quantity: Length (SI).
Definition: 1 stab = 1 metre.

Note: This name was proposed as an alternative for the metre, but never employed.

STAT St (O)

Quantity: Radioactive disintegration rate, i e activity (arbitary).
Definition: One stat is the quantity of radon that, due to its being situated in air, gives rise in one second to a charge of one franklin in air.

STAT-...

The names of units of unrationalised electrical and magnetic quantities in the CGS-esu system are obtained by prefixing the corresponding SI unit names by stat- (representing the words 'absolute electrostatic'). The following table contains single-word units only.

CGS-esu unit	unit symbol	quantity measured	corresponding CGSe unit of equal size	corresponding CGS-esu mechanical unit	size of one CGS-esu unit in SI units
statampère	statA	current	Fr/s	dyn$^{\frac12}$ cm/s	$10^{-1}/(c)$ ampère
statcoulomb	statC	charge	Fr	dyn$^{\frac12}$ cm	$10^{-1}/(c)$ coulomb
statfarad	statF	capacitance	Fr^2/erg	cm	$10^5/(c)^2$ farad
stathenry	statH	inductance	erg s^2/Fr^2	s^2/cm	$10^{-5}(c)^2$ henry
statohm	stat Ω	resistance; impedance; reactance	erg s/Fr^2	s/cm	$10^{-5}(c)^2$ ohm
statsiemens	statS	conductance; admittance; susceptance	Fr^2/erg s	cm/s	$10^5/(c)^2$ siemens
stattesla	statT	magnetic flux density	erg s/Fr cm^2	dyn$^{\frac12}$ s/cm^2	$10^{-2}(c)$ tesla
statvolt	statV	potential (electric); electromotive force	erg/Fr	dyn$^{\frac12}$	$10^{-6}(c)$ volt
statweber	statWb	magnetic flux	erg s/Fr	dyn$^{\frac12}$ s	$10^{-6}(c)$ weber

cm $\Rightarrow$ centimetre; dyn = dyne; Fr = franklin; s = second
(*c*) is the numerical value of the velocity of electromagnetic radiation in vacuo, expressed in metres per second: $(c) = 2{\cdot}997\,925 \times 10^8$

STATHM

Quantity: Mass (CGS).
Definition: 1 stathm = 1 gramme.
Note: This name (like the bes and the brieze) was proposed as an alternative for the gramme, but never employed.

STERADIAN

Quantity: (Geometrical) solid angle (all).
Symbol: sr. The unit symbol sterad has also been used.

Definition: One steradian is the solid angle which, having its vertex at the centre of a sphere, cuts off an area of surface equal to that of a square with sides of length equalling the radius of the sphere.
Note: A complete sphere contains 4π sr.

STÈRE

Quantity: Volume (SI).
Symbol: st; it was formerly s.
Definition: 1 st = 1 m³.
Note: This unit is now only used for the measurement of timber volumes.

STHÈNE   sn (BS)

Quantity: Force (MTS).
Definition: One sthène is the force which, when applied to a body of mass one tonne (t), gives it an acceleration of one metre per second squared (m/s²). 1 sn = 1 t m/s² = 1 000 newtons (N). 1 N = 0·001 sn.
Note: The unit was formerly called the funal.

STIGMA σ (O)

Quantity: Length, particularly nuclear lengths (metric).
Definition: 1 σ = 10^{-12} m. 1 m = 10^{12} σ.
Note: This unit has also been called the bicron.

STILB   sb

Quantity: Luminance (CGS).
Definition: 1 sb = 1 candela per square centimetre (cd/cm²) = 10^{4} nits (nt). 1 nt = 10^{-4} sb.

STOKES

Quantity: Kinematic viscosity (CGS).
Symbol: St; formerly it was S.
Definition: One stokes is the kinematic viscosity of a fluid with a dynamic viscosity of one poise (P) and a density of one gramme per centimetre cubed (g/cm³). 1 St = 1 P cm³/g = 1 (g/cm s)cm³/g = 1 cm²/s = 10^{-4} m²/s. 1 m²/s = 10^{4} St.
Note: The unit has also been called the stoke. It was formerly called the lentor.

STONE

Quantity: Mass (imperial).
Definition: 1 stone = 14 lb = 6·350 293 18 kg. 1 kg $\approx$ 0·157 473 stone.

Note: This is a UK unit; there is no corresponding US unit. See also the entry in Appendix 4.

STRONTIUM UNIT   SU (BS)

Quantity: Concentration of strontium-90 in an organic medium (eg soil, milk, bone) relative to the concentration of calcium in the same medium (arbitrary).
Definition: 1 SU = 10^{-9} curie of strontium-90 per kilogramme of calcium.
Note: This unit has also been called the sunshine unit.

STURGEON

Quantity: Magnetic reluctance (MKSA).
Definition: 1 sturgeon = 1 /international henry.
Note: This unit is obsolete.

SUNSHINE UNIT   SU (BS)

Quantity: Concentration of strontium-90 in an organic medium (eg soil, milk, bone) relative to the concentration of calcium in the same medium (arbitrary).
Definition: 1 SU = 10^{-9} curie of strontium-90 per kilogramme of calcium.
Note: The use of this name is deprecated. The unit is better called the strontium unit.

SVEDBERG   S (O)

Quantity: Sedimentation coefficient, ie time (all).
Definition: The sedimentation coefficient S in svedbergs is related to the velocity v of the boundary between the solution containing the molecules and solvent (in metres per second), the distance x from the boundary to the axis of rotation (in metres) and the angular velocity ω of the centrifuge causing sedimentation (in radians per second) by the formula $S = 10^{-13} \ v/\omega^2 x$. Ie 1 S = 10^{-13} second (s). 1 s = 10^{13} S.

TALBOT

Quantity: Luminous energy. (This is generally regarded as an SI unit, although it belongs to all systems.)
Definition: One talbot is the luminous energy corresponding to one joule (J) of radiant energy, having a luminous efficiency of one lumen per watt (lm/W). Since 1 W = 1 J/s, 1 talbot = 1 J lm/(J/s) = 1 lm s.
Note: The unit is equal in size to the lumberg (lumerg).

TECHNICAL UNIT OF MASS, BRITISH

See slug

TECHNICAL UNIT OF MASS, CGS-

See glug

TECHNICAL UNIT OF MASS, METRIC (SI)

See slug, metric

TENTHMETRE

Quantity: Length, especially the wavelength of visible and near-visible radiation (metric).
Definition: 1 tenthmetre $= 10^{-10}$ m. 1 m $= 10^{10}$ tenthmetres.
Note: This unit is better called the ångström.

TERA-...

See Appendix 3 section 2. Units are not entered under the prefix tera-.

TESLA   T

Quantity: Magnetic flux density (SI).
Definition: One tesla is the area-density of one weber (Wb) of magnetic flux per square metre (m²). 1 T $= 1$ Wb/m².
Note: This unit was formerly called the absolute tesla (T_{abs}), and must be distinguished from the international tesla (T_{int}), formally abandoned in 1948:
1 $T_{int} = 1 \cdot 000$ 34 T.   1 T $\approx 0 \cdot 999$ 660 T_{int}.

TEX

Quantity: Line density (metric).
Definition: One tex is the line density of a thread which has a mass of one gramme (g) and a length of one kilometre (km).
1 tex $= 1$ g/km $= 10^{-6}$ kg/m.   1 kg/m $= 10^{6}$ tex.
Note: This unit is used in the textile industry as a measure of yarn count.

THERM

Quantity: Heat energy (metric).
Definition: 1 therm $= 10^{5}$ (international table) British thermal units $= 1 \cdot 055$ 055 852 62 $\times 10^{8}$ joules (J).   1 J $\approx 9 \cdot 478$ 17 $\times 10^{-7}$ therm.

THERMIE   th (BS)

Quantity: Heat energy (arbitrary).
Definition: One thermie is the quantity of heat required to raise the temperature of one tonne (t) of air-free water from 14·5 °C to 15·5 °C at a constant pressure of one standard atmosphere. Since the fifteen-degree calorie (cal_{15}), referring to one gramme of water, is experimentally defined by 1 cal_{15} = 4·185 5 ± 0·000 5 joules (J) and 1 t = 1 000 kg = 10^6 g, 1 th = 10^6 cal_{15} ≈ 4·185 5 × 10^6 J. 1 J ≈ 2·389 2 th.

THOU

Quantity: Length (imperial).
Definition: 1 thou = 10^{-3} inch = 2·54 × 10^{-5} m. 1 m ≈ 3·937 01 × 10^4 thous.
Note: The thou is a popular name for one-thousandth of an inch (one milli-inch). It is also popularly called the mil.

THOUSANDTH MASS UNIT   TMU (O)

Quantity: Energy (arbitrary).
Definition: One thousandth mass unit is the energy equivalent of the mass of a hypothetical molecule with a relative molecular mass of 0·001 gramme per mole (physical scale). Using Einstein's equation $E = mc^2$, with m = 1 atomic mass unit (physical) and c = velocity of electromagnetic radiation, the experimentally derived value is 1 TMU ≈ 1·491 76 × 10^{-13} joule (J). 1 J ≈ 6·703 49 × 10^{12} TMU.

TOLERANCE UNIT

Quantity: Engineering tolerances (none).
Definition: If D is the geometrical mean of the diameter steps involved, the size of the tolerance unit i is given by:
   $i = 0·45 D^{\frac{1}{3}} + 0·001 D$ (D in millimetres, i in micrometres);
   $i = 0·052 D^{\frac{1}{3}} + 0·001 D$ (D in inches, i in milli-inches).

TON (1)

Quantity: Explosive power of a nuclear weapon (arbitrary).
Definition: A nuclear weapon of power one ton has an energy equivalent of one ton of trinitrotoluene. 1 ton = 5 × 10^9 joules approximately.
Note: The units normally employed are the kiloton (10^3 ton) and the megaton (10^6 ton).

TON (2), (3)

Quantity: Mass (imperial).
(2) ˙Avoirdupois measure
    Definition: 1 ton = 20 hundredweights = 2 240 pounds = 1 016·046 908 8 kg. 1 kg ≈ 9·842 07 × 10^{-4} ton.
    Note: (1) This is the UK name for the unit. In the US, where it is almost obsolete, it is called the long ton or the gross ton. The US unit called the ton is more properly called the short ton or the net ton.
    (2) The name displacement ton is employed when referring to the tonnage of ships.
(3) Troy measure   ton tr (O)
    Definition: 1 ton tr = 20 troy hundredweights = 2 000 troy pounds = 746·483 443 2 kg. 1 kg ≈ 0·001 339 62 ton tr.
Note: (3) See Appendix 4 for measurement ton, register ton, shipping ton.

TON, ASSAY

Quantity: Mass (metric).
Definition: (1) One UK assay ton contains as many milligrammes (mg) as a (long) ton contains troy ounces.
    1 UK assay ton = (2 240 × 7 000)/480 mg ≈ 0·032 666 7 kg.
    1 kg ≈ 30·612 2 UK assay tons.
(2) One US assay ton contains as many milligrammes as a short ton contains troy ounces.
    1 US assay ton = (2 000 × 7 000)/480 mg ≈ 0·029 166 7 kg.
    1 kg ≈ 34·285 7 US assay tons.
Note: 1 UK assay ton = 1·12 US assay ton; 1 US assay ton ≈ 0·892 857 UK assay ton. The US unit is sometimes called the short assay ton.

TONDAL

Quantity: Force (imperial).
Definition: One tondal is the force which, when applied to a body of mass one ton, gives it an acceleration of one foot per second squared (ft/s²). 1 tondal = 1 ton ft/s² = 309·691 097 802 24 newtons (N).
1 N ≈ 0·003 229 02 tondal.

TON-FORCE   tonf (BS)

Quantity: Force (imperial).
Definition: One ton-force is the force which, when applied to a body of mass one ton, gives it an acceleration equal to the standard acceleration of free fall (g_n). Since g_n = 9·806 65 metres per second squared = 9·806 65/0·304 8 feet per second squared (ft/s²) ≈ 32·174 0 ft/s², 1 tonf = 9 964·016 418 183 52 newtons (N). 1 N ≈ 1·003 61 × 10^{-4} tonf.

TONNE = TONNEAU   t

Quantity: Mass (MTS).
Definition: 1 t = 1 000 kg. 1 kg = 0·001 t.
Note: This unit is also called the metric ton and the millier.

TON OF REFRIGERATION

Quantity: Refrigerating capacity, ie heat flow rate (power) (imperial).
Definition: One ton of refrigeration is the rate of extraction of heat when one short ton (sh ton) of ice of specific latent heat 144 international table British thermal units per pound (Bt_{uIT}/lb) is produced in 24 hours (h) from water at the same temperature. Since 1 sh ton = 2 000 lb and 24 h = 1 day, 1 ton of refrigeration = 2 000 × 144 Bt_{uIT}/day. Also, since 1 sh ton = 907·184 74 kg, 1 Bt_{uIT}/lb = 2 326 joules per kilogramme (J/kg) and 24 h = 86 400 s, 1 ton of refrigeration = (907·184 74 × 144 × 2 326)/86 400 J/s (watt, W) ≈ 3 516·85 W. 1 W ≈ 2·843 45 × 10^{-4} ton of refrigeration.
Note: The shortened name ton is often used.

TON, SHORT

Quantity: Mass (imperial).
Symbol: sh ton (BS); in the US, sh tn.
Definition: 1 sh ton = 2 000 lb = 907·184 74 kg. 1 kg ≈ 0·001 102 31 sh ton.
Note: This is a US unit, sometimes called net ton or just ton. There is no corresponding UK unit.

TON-WEIGHT   tonwt

Quantity: Force (imperial).
Definition: One ton-weight is the force which, when applied to a body of mass one ton, gives it an acceleration equal to the local value of the acceleration of free fall (g) expressed in feet per second squared.
1 tonwt = 309·691 097 802 24 g newton (N). 1 N ≈ 0·003 229 02/g tonwt.
Note: This is an inconsistent unit and its use is deprecated. The ton-force should always be used in its place.

TOR

Quantity: Pressure (SI).
Definition: One tor is the pressure resulting from a force of one newton (N) acting uniformly over an area of one square metre (m^2).
1 tor = 1 N/m^2.
Note: This unit is now called the pascal. It must not be confused with the torr.

TORR

Quantity: Pressure (metric).
Definition: 1 torr = $\frac{1}{760}$ atmosphere $\approx$ 133·322 pascals (Pa).
1 Pa $\approx$ 0·007 500 62 torr.
Note: (1) The torr is equal in size to the (conventional) millimetre of mercury to one part in seven million.
(2) The torr must not be confused with the tor.

TRANSMISSION UNIT

Quantity: Intensity level (all).
Definition: 1 transmission unit = 1 decibel.
Note: This was one of several names proposed as alternatives for the decibel, but almost never employed.

TROLAND

Quantity: Retinal illumination (metric).
Definition: One troland is the retinal illumination produced by a surface having a luminance of one nit when the area of the pupil of the eye is one square millimetre.
Note: The unit has also been called the luxon. It was originally called the photon.

TSI

Quantity: Pressure (imperial).
Definition: One tsi is the pressure resulting from a force of one ton-force (tonf) acting uniformly over an area of one square inch (in²).
1 tsi = 1 tonf/in² $\approx$ 1·544 43 $\times$ 10⁷ pascals (Pa). 1 Pa $\approx$ 6·474 90 $\times$ 10⁻⁸ tsi.

TURN   T (O)

Quantity: (Geometrical) plane angle (all).
Definition: 1 T = 2π rad = 360°.
Note: This unit is also called the circle.

TYPP

Quantity: Reciprocal line density (imperial).
Definition: One typp is the reciprocal line density of a thread which has a length of one thousand yards (yd) and a mass of one pound (lb).
1 typp = 1 000 yd/lb $\approx$ 2 015·91 m/kg. 1 m/kg $\approx$ 0·000 496 055 typp.
Note: This unit is used in the textile industry as an indirect measure of yarn count. The direct unit of yarn count the tex is used in preference to the typp.

VAC

Quantity: Pressure (metric).
Definition: 1 vac = 100 pascals (Pa). 1 Pa = 0·01 vac.
Note: 1 vac = 1 millibar.

VAR

Quantity: Reactive power of an alternating current (SI).
Definition: One var is the product of one volt (V) and one ampère (A).
1 var = 1 V A = 1 watt (W).

VIBRATION   vib (O)

Quantity: Frequency (all).
Definition: One vibration is the frequency of a periodic occurrence which has a period of one second (s).
1 vib = 1 /s. 1 s = 1 /vib.
Note: (1) This unit is better called the hertz. It is also called the cycle (per second).
(2) In France the vibration is used with the following meaning:
1 vibration (French) = 0·5 vib = 0·5 hertz.

VIOLLE

Quantity: Luminous intensity (metric).
Definition: One violle is the luminous intensity of one square centimetre of platinum at its temperature of solidification.
1 violle = 20·17 candelas (measured comparison).
Note: The temperature of solidification of platinum is 2 045 K.

VOLT   V

Quantity: Electric potential and potential difference, electromotive force (SI).
Definition: One volt is the difference in potential between two points of a conducting wire carrying a constant current of one ampère (A), when the power dissipated between these points is equal to one watt (W).
1 V = 1 W/A.
Note: (1) Several former definitions of the volt exist:
(a) One volt is (approximately) the electromotive force of a Daniell cell. This was used from 1860 by Lord Kelvin, and recorded by the British Association for the Advancement of Science in 1873. The present-day accepted value for the electromotive force of the Daniell cell is 1·08 V.
(b) In 1893 the Fourth International Electrical Congress defined the international volt (V_{int}) as follows: One international volt is represented with

sufficient accuracy as 1 000/1 434 of the electromotive force of a Latimer-Clark cell. This definition became law in the UK and the US in 1894, in France (by decree) in 1896, and in Germany in 1898. The present-day accepted value for the electromotive force E_θ(L–C) of the Latimer-Clark cell at a temperature θ °C is given in volts by:

E_θ (L–C) = 1·433 3 − 0·001 19 $(\theta - 15)$ − 7 × 10^{-6} $(\theta - 15)^2$.

The present-day volt was described as an absolute volt (V_{abs}) to distinguish it from the international volt.

(c) Later the Latimer-Clark cell was replaced by the Weston (cadmium) cell at 20 °C. The electrolyte is a saturated solution, and excess $CdSO_4 . \frac{8}{3} H_2O$ crystals are always present; decinormal sulphuric acid is used. The electromotive force E_θ(W) at a temperature θ °C is given in volts by:
E_θ(W) =

$$1·018 58 − 4·06 \times 10^{-5}(\theta - 20) − 9·5 \times 10^{-7}(\theta - 20)^2 + 1 \times 10^{-8}(\theta - 20)^3.$$

(2) The relationship between the (absolute) volt and the international volt has now been formally defined as 1 V_{int} = 1·000 34 V. 1 V ≈ 0·999 660 V_{int}. An early name for the international volt is the ohma.

VOLTAMPERE   VA

Quantity: Apparent power of an alternating current (SI).
Definition: One voltampère is the product of one volt (V) and one ampère (A). 1 VA = 1 V A = 1 watt (W).

VOLT, EQUIVALENT

Quantity: Energy, particularly in atomic studies (arbitrary).
Definition: 1 equivalent volt = 1 electronvolt.
Note: This was the former name for the electronvolt.

VOLT, THERMAL

Quantity: Thermal potential difference (all).
Definition: One thermal volt across a conductor of heat corresponds to a temperature difference of one kelvin (K). 1 thermal volt = 1 K.

VOLUME UNIT

Quantity: Magnitude of a complex electric wave, eg corresponding to speech or music (all).
Definition: One volume unit is equal to the number of decibels by which the wave differs from a reference magnitude.

WATT W

Quantity: Power (SI).
Definition: One watt is the power available when energy of one joule (J) is expended in one second (s). 1 W = 1 J/s.
Note: (1) This unit was formerly called the absolute watt (W_{abs}), and must be distinguished from the (mean) international watt (W_{int}), formally abandoned in 1948:
1 W_{int} = $(1 \cdot 000\ 34)^2/1 \cdot 000\ 49$ W $\approx 1 \cdot 000\ 19$ W. 1 W $\approx 0 \cdot 999\ 810\ W_{int}$.
(2) The var and the voltampère are units each identical with the watt, but used for special kinds of electrical power. The former definition of the thermal ampère made it equal to the watt.

WEBER Wb (1)

Quantity: Magnetic flux (SI).
Definition: One weber is the magnetic flux which, linking a circuit of one turn, produces in it an electromotive force of one volt (V) as it is reduced to zero at a uniform rate in one second (s). 1 Wb = 1 V s.
Note: (1) This unit was formerly called the absolute weber (Wb_{abs}), and must be distinguished from the international weber (Wb_{int}), formally abandoned in 1948:
1 Wb_{int} = $1 \cdot 000\ 34$ Wb; 1 Wb $\approx 0 \cdot 999\ 660\ Wb_{int}$.
(2) At one time the weber, like the galvat, was used as the international unit of current; it was soon replaced by the international ampère.

WEBER Wb (2)

Quantity: Magnetic pole strength (CGS-emu).
Definition: 1 Wb = 1 dyne per oersted.
Note: There is no SI equivalent because pole strength is not an SI quantity.

X-UNIT

Quantity: Length, particularly wavelength in X-ray spectra (arbitrary).
Symbol: X. The unit symbol XU has also been used.
Definition: The experimentally derived value was subsequently formalised by assigning the value $3\ 029 \cdot 45$ X to the spacing of the (200) planes of calcite at 18 °C. This gives:
1 X = $(1 \cdot 002\ 02 \pm 0 \cdot 000\ 03) \times 10^{-13}$ m. 1 m $\approx 9 \cdot 979\ 84 \times 10^{12}$ X.
Note: The unit is also called the X-ray unit; alternatively it is called the Siegbahn unit. It was abandoned in 1948 in favour of the ångström, which was itself subsequently abandoned in favour of the nanometre.

YARD

Quantity: Length (imperial).
Symbol: yd. The sign $^\times$ is sometimes used, but this is not recommended.
Definition: 1 yd = 0·914 4 m. 1 m $\approx$ 1·093 61 yd.
Note: (1) The definition has been given legal enforcement in the UK in the Weights and Measures Act of 1963, and in the US in Document 59–5442 of the US Department of Commerce, National Bureau of Standards, 1959.
(2) The UK yard was originally defined in the Weights and Measures Act of 1878 as follows: The straight line or distance between the centres of the two gold plugs or pins (as mentioned in the First Schedule to this Act) in the bronze bar by this Act declared to be the imperial standard for determining the imperial standard yard measured when the bar is at the temperature of sixty-two degrees of Fahrenheit's thermometer, and when it is supported on bronze rollers placed under it in such a manner as best to avoid flexure of the bar, and to facilitate its free expansion and contraction from variations of temperature, shall be the legal standard measure of length, and shall be called the imperial standard yard, and shall be the only unit or standard measure of extension from which all other measures of extension, whether linear, superficial or solid, shall be ascertained. In 1922 the UK yard as defined by this Act was compared experimentally with the metre: the result was 1 UKyd = 36/39·370 147 m $\approx$ 0·914 398 41 m. In 1960 the bronze bar was found to be contracting at a rate of approximately 10^{-6} inch per annum.
(3) The US yard was originally defined in the US law of 1866, and made definitive in the Mendenhall Order of 5 April 1893. Its size was given by 1 USyd = 36/39·37 m; ie 1 USyd $\approx$ 0·914 401 83 m.
(4) See Appendix 4 for cubic yard, yard of land.

YEAR

Quantity: Time (arbitrary).
Symbol: y(O) or a.
Definition: There are several types of year. The values for their lengths in days have been experimentally determined; the other values have been computed from them.

type of year	d	d	h	m	s	10^7 s
anomalistic	365·259 641 34 + 3·04 × 10^{-6}T	365	06	13	53·012	3·155 843 301 2
Gaussian	365·256 898	365	06	09	55·8	3·155 819 58
Gregorian	365·242 5	365	05	49	12	3·155 695 2
Julian	365·25	365	06			3·155 76
nodical	346·620 031 + 3·2 × 10^{-5}T	346	14	52	50·7	2·994 797 07
sidereal	365·256 365 56 + 1·1 × 10^{-7}T	365	06	09	09·984	3·155 814 998 4
tropical	365·242 198 781 − 6·14 × 10^{-6}T	365	05	48	45·9747	3·155 692 59747
360°	365·255 189 7	365	06	07	28·39	3·155 804 839

All times are related to the ephemeris second. Thus, in the column in day units, the change in the length of the year is given (where appropriate and known); T is the number of Julian centuries (of 365 25 d) after 1900 January 0 at 12 h ephemeris time. The corresponding variations are not shown in the other columns.

---

anomalistic year:	time interval between two consecutive passages of the earth through perihelion
Gaussian year:	time period derived from Kepler's laws, based on the earth-sun distance being one astronomical unit
Gregorian year:	calendar year with 97 intercalated days in every 400 years
Julian year:	calendar year with one intercalated day in every four years
nodical year:	(also called draconic or eclipse year) time interval between two consecutive passages of the earth through the ascending (or descending) node of its orbit
sidereal year:	time period during which the earth returns to the same point in its orbit with reference to the stars
tropical year:	time interval between two consecutive passages, in the same direction, of the sun through the earth's equatorial plane
360° year:	time period for a 360° revolution of the mean sun, measured from a fixed equinox

---

PART 2
PHYSICAL QUANTITIES

ABSORPTION AREA, EQUIVALENT E (O)

Definition: That surface area having a reverberation absorption coefficient α of unity which would absorb sound energy at the same rate as the room or object in the room. $E = \alpha A$ (A = surface area of the room or object).
Unit: Metre squared. L^2 (scalar).
Note: The equivalent absorption area depends on the frequency of the sound.

ABSORPTION COEFFICIENT (1)

Symbol: α; occasionally α_a.
Definition: Energy E_a absorbed by a surface divided by that incident on it (E_o) under identical conditions. $\alpha = E_a/E_o$.
Unit: There are no units or dimensions (scalar).
Note: (1) The quantity is also termed the absorption factor.
(2) The absorption coefficient is generally expressed as a percentage.
(3) $\alpha = \delta + \tau$, where δ = scattering coefficient and τ = transmission coefficient.
(4) In sound, the absorption coefficient depends on the frequency of the sound. When the incident sound is distributed completely at random the quantity is termed the reverberation absorption coefficient.

ABSORPTION COEFFICIENT μ_α (BS) (2)

Definition: For a parallel beam of radiation passing through a uniform medium, the absorption coefficient is the factor μ_α in the expression $e^{-\mu_\alpha x}$ for the fraction remaining unabsorbed after passing through a layer x.
Unit: (1) Linear absorption coefficient (x in metres)
    Per metre. L^{-1} (vector).
(2) Mass absorption coefficient (x in kilogrammes per metre squared)
    Metre squared per kilogramme. $M^{-1} L^2$ (vector).
(3) Molar absorption coefficient (x in moles per metre squared)
    Metre squared per mole. $L^2 N^{-1}$ (vector).
Note: The absorption coefficient is less than the attenuation coefficient since it does not include that part of the energy which escapes in the form of secondary radiation.

ABSORPTIVITY

Definition: The internal absorption coefficient of unit thickness of an absorbing material under conditions in which the boundary of the material has no influence.
Unit: There are no units or dimensions (scalar).

ACCELERATION

Symbol: a. The symbol f, formerly used, is no longer preferred.
Definition: Rate of change of velocity v with time t. $a = dv/dt$.
Unit: Metre per second squared. L T^{-2} (vector).
Note: The symbol g is used for the value of the local acceleration of free fall; the standard value g_n is given by $g_n = 9 \cdot 806\ 65$ metres per second squared.

ACCELERATION, ANGULAR

Symbol: ϕ; sometimes a.
Definition: Rate of change of angular velocity ω with respect to time t.
$\phi = d\omega/dt$.
Unit: Radian per second squared, degree per second squared, hertz squared. T^{-2} (vector).

ACTION L (O)

Definition: The integral of energy E with respect to time t. $L = \int E\ dt$.
Unit: Joule second. M L^2 T^{-1} (scalar).

ACTIVITY

Definition: The number N of disintegrations taking place in a radioactive specimen divided by the time t. Activity $= N/t$.
Unit: Curie, rutherford. T^{-1} (scalar).
Note: (1) The quantity is also termed the radioactive disintegration rate.
(2) See also the entry power (note 1).

ADMITTANCE Y

Definition: The reciprocal of the electrical impedance Z. $Y = 1/Z$.
Unit: Siemens. M^{-1} L^{-2} T^3 I^2; L^{-1} T μ^{-1}; L T^{-1} ε (scalar).
Note: $Y = G + iB$ (G = conductance, B = susceptance). $Y = |Y| e^{-i\phi}$ (ϕ = phase displacement); hence $|Y| = (G^2 + B^2)^{\frac{1}{2}}$. The term admittance is sometimes given to the modulus $|Y|$, Y being termed the complex admittance.

ALTITUDE

See length, note

AMPÈRAGE

See current (electric), note

AMPLITUDE LEVEL N (O)

Definition: The natural logarithm of the ratio of two amplitudes a_1, a_2, each expressed in the same units. $N = \ln(a_2/a_1)$.
Unit: Neper. There are no dimensions (scalar).
Note: The name and symbol given are commonly used for this quantity, but neither is standard.

ANGLE (PLANE)

Symbol: θ, a. Other convenient lowercase letters of the Greek alphabet are also employed, eg β, γ, ϕ.
Definition: The arc length s cut out on a circle divided by the radius r of the circle. $\theta = s/r$.
Unit: Radian, degree. There are no dimensions (scalar, although a direction (eg clockwise or anticlockwise) can if necessary be associated with the angle).
Note: (1) The quantity defined as above was formerly more correctly called analytical angle, and possessed no units. The size of an analytical angle was made equal to the size of the corresponding geometrical angle, the last named being defined as the region cut out in a plane between two straight lines terminating in the same point and measured in radians.
(2) The phase angle (phase displacement, phase difference) is the constant ϕ (or θ) in the equation for a variable x which alters with time t, $x = \hat{x} \sin(\omega t + \phi)$, ω being another constant.

ANGLE, SOLID

Symbol: ω. Sometimes Ω is used.
Definition: The area A cut out on a spherical surface divided by the square of the radius r of the sphere. $\omega = A/r^2$.
Unit: Steradian. There are no dimensions (scalar).
Note: The quantity defined as above was formerly more correctly called analytical solid angle, and possessed no units. The size of an analytical solid angle was made equal to the size of the corresponding geometrical solid angle, the last named being defined as the region cut out in space by an arbitrary cone and measured in steradians.

APERTURE CONDUCTIVITY a (BS)

Definition: Density ρ of a medium divided by acoustical mass m_a at the aperture. $a = \rho/m_a$.
Unit: Kilogramme per pascal second squared. L (scalar).

AREA

Symbol: A; sometimes S.
Definition: The region of a surface contained by a closed curve. The size of the area will be expressed by a formula, the nature of which depends on the shape of the region.
Unit: Metre squared. L^2 (vector).
Note: Surface area refers to the region that separates a solid from its surroundings.

AREA MOMENT, MAGNETIC

See moment, electromagnetic (note)

AREA, SPECIFIC   *a*

Definition: Area A divided by mass m. $a = A/m$.
Unit: Metre squared per kilogramme. $M^{-1} L^2$ (vector)

ATTENUATION COEFFICIENT (1)

Symbol: α or μ(BS); sometimes a.
Definition: For a parallel beam of radiation passing through a uniform medium, the attenuation coefficient is the factor α in the expression $e^{-\alpha x} \cos \beta (x - x_0)$ for the fraction remaining unattenuated after passing through a layer x. β (or b) is the phase coefficient. The propagation coefficient γ (or p) is defined by $\gamma = \alpha + i\beta$. If $\alpha \approx 0$, β is often replaced by the circular wave number.
Unit: (1) Linear attenuation coefficient (x in metres)
   (Neper) per metre. L^{-1} (vector).
(2) Mass attenuation coefficient (x in kilogrammes per metre squared)
   Metre squared per kilogramme. $M^{-1} L^2$ (vector).
(3) Molar attenuation coefficient (x in moles per metre squared)
   Metre squared per mole. $L^2 N^{-1}$ (vector).
Note: The attenuation coefficient is greater than the absorption coefficient since it takes account of secondary radiation.

ATTENUATION COEFFICIENT (2)

Definition: The sum of the scattering coefficient δ and the absorptivity a. Attenuation coefficient $= \delta + a$.
Unit: There are no units or dimensions (scalar).
Note: The quantity is also termed the attenuation factor.

BREADTH

See length (note)

BRIGHTNESS

See luminance (note)

BRIGHTNESS, APPARARENT (or SUBJECTIVE)

See luminosity (note)

BRILLIANCE, POINT

Definition: The illumination produced by a source on a plane at the observer's eye normal to the incident flux.
Unit: Lux. L^{-2} F; M T^{-3} (scalar).
Note: The quantity refers to a source so distant that its apparent diameter is insignificant.

CALORIFIC VALUE, VOLUME BASIS C (O)

Definition: Heat energy Q divided by volume V. $C = Q/V$.
Unit: Joule per metre cubed. L^{-3} Q; M L^{-1} T^{-2} (scalar).

CANDLE POWER

See intensity, luminous (note 1)

CAPACITANCE C

Definition: Charge Q divided by potential (difference) V. $C = Q/V$.
Unit: Farad. M^{-1} L^{-2} T^4 I^2; L^{-1} T^2 μ^{-1}; L ε (scalar).

CAPACITANCE, SPECIFIC INDUCTIVE

See permittivity, relative (note)

CAPACITANCE, THERMAL

Definition: Entropy S divided by temperature difference θ. Thermal capacitance $= S/\theta$.
Unit: Thermal farad $=$ joule per kelvin squared. Q Θ^{-2}; M L^{-2} T^2 (scalar).
Note: The former definition was: heat energy Q divided by temperature difference. Thermal capacitance $= Q/\theta$.

CAPACITIVITY

See permittivity, absolute (note 1)

CAPACITY

See volume (note)

CHARGE Q

Definition: The integral of current I with respect to time t. $Q = \int I \, dt$.
Unit: Coulomb. T I; $M^{\frac{1}{2}} L^{\frac{1}{2}} \mu^{-\frac{1}{2}}$; $M^{\frac{1}{2}} L^{3/2} T^{-1} \varepsilon^{\frac{1}{2}}$ (scalar).
Note: The quantity is sometimes inadequately termed the quantity of electricity.

CHARGE DENSITY

Symbol: ρ; sometimes η.
Definition: Charge Q divided by volume V. $\rho = Q/V$.
Unit: Coulomb per metre cubed. $L^{-3} T I$; $M^{\frac{1}{2}} L^{-5/2} \mu^{-\frac{1}{2}}$; $M^{\frac{1}{2}} L^{-3/2} T^{-1} \varepsilon^{\frac{1}{2}}$ (scalar).
Note: The quantity is better termed the volume density of charge.

CHARGE DENSITY, SURFACE σ

Definition: Charge Q divided by surface area A. $\sigma = Q/A$.
Unit: Coulomb per metre squared. $L^{-2} T I$; $M^{\frac{1}{2}} L^{-3/2} \mu^{-\frac{1}{2}}$; $M^{\frac{1}{2}} L^{-\frac{1}{2}} T^{-1} \varepsilon^{\frac{1}{2}}$ (scalar).

CHARGE, SPECIFIC q (O)

Definition: Charge Q divided by mass m. $q = Q/m$.
Unit: Coulomb per kilogramme. $M^{-1} T I$; $M^{-\frac{1}{2}} L^{\frac{1}{2}} \mu^{-\frac{1}{2}}$; $M^{-\frac{1}{2}} L^{3/2} T^{-1} \varepsilon^{\frac{1}{2}}$ (scalar).

CHARGE, THERMAL

See entropy (note)

COMPLIANCE, ACOUSTICAL C_a (BS)

Definition: The reciprocal of the acoustical stiffness S_a. $C_a = 1/S_a$.
Unit: Metre cubed per pascal. $M^{-1} L^4 T^2$ (vector).

COMPLIANCE (MECHANICAL) C

Definition: The reciprocal of the mechanical stiffness s. $C = 1/s$.
Unit: Metre per newton. $M^{-1} T^2$ (vector).

COMPRESSIBILITY (BULK)

Symbol: χ; sometimes κ.
Definition: The reciprocal of the bulk modulus of elasticity K. $\chi = 1/K$.
Alternatively, the compressibility can be defined by the equation $\chi = \dfrac{1}{V}\dfrac{dV}{dp}$
under specified conditions (V = volume, p = pressure); the condition
specified is generally constant temperature.
Unit: Per pascal = metre squared per newton. $M^{-1} L T^2$ (scalar).
Note: The quantity is more correctly termed the coefficient of compressibility.

COMPRESSION, MODULUS OF

See elastic modulus (3)

CONCENTRATION

See density (note)

CONDUCTANCE G

(1) CONDUCTANCE TO A DIRECT CURRENT
    Definition: The reciprocal of the electrical resistance R. $G = 1/R$.
(2) CONDUCTANCE TO AN ALTERNATING CURRENT
    Definition: The real part of the admittance Y. $Y = G + iB$
    (B = susceptance).
Unit: Siemens. $M^{-1} L^{-2} T^3 I^2$; $L^{-1} T \mu^{-1}$; $L T^{-1} \varepsilon$ (scalar).

CONDUCTANCE, SPECIFIC

See conductivity, electrical (note)

CONDUCTANCE, THERMAL

Symbol: h; sometimes K or U or a. (K or a are preferred for surface thermal
conductance.)
Definition: Heat flow rate density q divided by temperature difference θ.
$h = q/\theta$.
Unit: Watt per metre squared kelvin. $L^{-2} T^{-1} Q \Theta^{-1}$; $M L^{-2} T^{-1}$ (scalar).
Note: The quantity is also termed the heat transfer coefficient or thermal
transmittance.

CONDUCTIVITY, ELECTRICAL

Symbol: γ or σ.
Definition: The reciprocal of the resistivity ρ. $\gamma = 1/\rho$.
Unit: Siemens per metre. $M^{-1} L^{-3} T^3 I^2$; $L^{-2} T \mu^{-1}$; $T^{-1} \varepsilon$ (scalar).
Note: The quantity is also termed the specific conductance.

CONDUCTIVITY, THERMAL

Symbol: λ; sometimes k.
Definition: Quantity of heat Q flowing in time t normally between two areas A a distance x apart and differing in temperature by θ, when the system is in a steady state. $\dfrac{dQ}{dt} = -\lambda A \dfrac{d\theta}{dx}.$ (The minus sign is accounted for by defining λ as positive; $d\theta/dx$ is negative for positive dQ/dt.)
Unit: Watt per metre kelvin. $L^{-1} T^{-1} Q \Theta^{-1}$; $M L^{-1} T^{-1}$ (scalar).
Note: The equivalent thermal conductivity λ_e of a body (such as a wall made up of slabs of thicknesses $x_1, x_2, \ldots$ and thermal conductivities $\lambda_1, \lambda_2, \ldots$ is given by $\lambda_e = \Sigma x_i / \Sigma(x_i/\lambda_i)$.

CONSTANT, DIELECTRIC

See permittivity, relative (note)

CONSTANT, ELECTRIC

See permittivity, absolute (note 1)

CONSTANT, MAGNETIC

See permeability, absolute (note 1)

COUPLE

See moment (note)

COUPLING COEFFICIENT

Symbol: k or κ.
Definition: Mutual inductance M divided by the geometrical mean of the self inductances L_1, L_2. $k = M/\sqrt{(L_1 L_2)}$.
Unit: There are no units or dimensions (scalar).
Note: The leakage coefficient (σ) is one minus the square of the coupling coefficient. $\sigma = 1 - k^2 = 1 - M^2/L_1 L_2$.

CURRENT DENSITY

Symbol: J; occasionally S.
Definition: The integral of the current density with respect to area A is equal to the current I. $I = \int J \, dA$.
Unit: Ampère per metre squared. L^{-2} I; $M^{\frac{1}{2}} L^{-3/2} T^{-1} \mu^{-\frac{1}{2}}$; $M^{\frac{1}{2}} L^{-\frac{1}{2}} T^{-2} \varepsilon^{\frac{1}{2}}$ (vector).

CURRENT DENSITY, LINEAR

Symbol: A; sometimes a.
Definition: Current I divided by breadth b of the conductor. $A = I/b$.
Unit: Ampère per metre. L^{-1} I; $M^{\frac{1}{2}} L^{-\frac{1}{2}} T^{-1} \mu^{-\frac{1}{2}}$; $M^{\frac{1}{2}} L^{\frac{1}{2}} T^{-2} \varepsilon^{\frac{1}{2}}$ (vector).

CURRENT (ELECTRIC) *I*

Definition: There is no formal definition of current. It can only be explained in terms of an inadequate synonym, eg flow of electric charge.
Unit: Ampère. I; $M^{\frac{1}{2}} L^{\frac{1}{2}} T^{-1} \mu^{-\frac{1}{2}}$; $M^{\frac{1}{2}} L^{3/2} T^{-2} \varepsilon^{\frac{1}{2}}$ (scalar).
Note: Engineers sometimes term the quantity amperage.

CURRENT, THERMAL

Definition: The rate of flow of entropy S with respect to time t.
Thermal current $= dS/dt$.
Unit: Thermal ampère = watt per kelvin. $T^{-1} Q \Theta^{-1}$; $M T^{-1}$ (scalar).
Note: The former definition was the rate of flow of heat energy Q with respect to time. Thermal current $= dQ/dt$.

CURVATURE *C* (O)

Definition: The reciprocal of the radius of curvature r. $C = 1/r$.

In cartesian co ordinates, $C = \dfrac{d^2y}{dx^2} \Big/ \left[1 + \left(\dfrac{dy}{dx}\right)^2\right]^{3/2}$.

Unit: Metre to the power of minus one. L^{-1} (vector).

DAMPING COEFFICIENT

Symbol: δ; occasionally Δ.
Definition: The quantity δ in the function f(t) defined by

$$\mathrm{f}(t) = A\, e^{-\delta t} \sin \frac{2\pi t - t_o}{T}.$$

Unit: (Neper) per second. T^{-1} (scalar).
Note: The quantity is also termed decay coefficient or decay factor.

DECAY COEFFICIENT

See damping coefficient (note)

DENSITY ρ

Definition: Mass m divided by volume V. $\rho = m/V$.
Unit: Kilogramme per metre cubed. M L^{-3} (scalar).
Note: The quantity is more properly called mass density. It is called concentration when it refers to the mass of one substance dissolved in a volume of a second substance.

DENSITY, AREA

Definition: Mass m divided by area A. Area density $= m/A$.
Unit: Kilogramme per metre squared. M L^{-2} (scalar).

DENSITY, LINE ρ' (O)

Definition: Mass m divided by length l. $\rho' = m/l$. Alternatively, for a body of uniform cross section, the line density is the product of its density ρ and cross-sectional area A. $\rho' = \rho A$.
Unit: Kilogramme per metre. M L^{-1} (scalar).

DENSITY, OPTICAL

See transmission coefficient (note 5)

DENSITY, RELATIVE

Symbol: d. The BS symbol is Sθ_1/θ_2, where θ_1, θ_2 are the temperatures of the two substances, θ_2 being that of the reference substance.
Definition: Density ρ_1 of a substance divided by the density ρ_2 of a reference substance under conditions specified for both substances (normally equal temperatures and pressures). $d = \rho_1/\rho_2$.
Unit: There are no units or dimensions (scalar).
Note: (1) The quantity has also been called specific gravity when the reference substance is water. For gases, it is also called vapour density when the reference substance is hydrogen.
(2) The specified temperatures are often $\theta_1 = \theta_2 = 20$ °C. At 3·98 °C the density of water is 1 000 kilogrammes per metre cubed, and thus the specific gravity of a substance is numerically equal to one-thousandth of its density.

DIFFUSION COEFFICIENT D (O)

Definition: Mass m of substance transported in time t across an area A when there is a concentration gradient dc/dz across this area.

$$\frac{dm}{dt} = DA\frac{dc}{dz}$$

Unit: Metre squared per second. $L^2\ T^{-1}$ (scalar).
Note: When the above equation refers to the diffusion of a solute through a solution it is known as Fick's law.

DIFFUSIVITY, THERMAL

Symbol: a; occasionally κ or a.
Definition: Thermal conductivity λ divided by the product of the density ρ and the specific heat capacity at constant pressure c_p. $a = \lambda/\rho c_p$.
Unit: Metre squared per second. $L^2\ T^{-1}$ (scalar).

DIPOLE MOMENT, ELECTRIC

Symbol: p; occasionally p_e.
Definition: The vector product of the dipole moment and the electric field strength E is equal to the moment of the force M. $p \times E = M$.
Unit: Coulomb metre. $L\ T\ I$; $M^{\frac{1}{2}}\ L^{3/2}\ \mu^{-\frac{1}{2}}$; $M^{\frac{1}{2}}\ L^{5/2}\ T^{-1}\ \varepsilon^{\frac{1}{2}}$ (vector).

DIPOLE MOMENT, MAGNETIC j

Definition: The vector product of the magnetic dipole moment and the magnetic field strength H is equal to the moment of the force M.
$j \times H = M$.
Unit: Weber metre. $M\ L^3\ T^{-2}\ I^{-1}$; $M^{\frac{1}{2}}\ L^{5/2}\ T^{-1}\ \mu^{\frac{1}{2}}$; $M^{\frac{1}{2}}\ L^{3/2}\ \varepsilon^{-\frac{1}{2}}$ (vector).
Note: (1) The quantity was formerly termed the magnetic moment, a name now reserved as an alternative for the electromagnetic moment.
(2) $(U/R)\overset{*}{j} = j/4\pi$; $\overset{*}{j} \times \overset{*}{H} = M$.

DISINTEGRATION RATE, RADIOACTIVE

See activity (note 1)

DISPLACEMENT

See length (note)

DISPLACEMENT, ELECTRIC D

Definition: The divergence of the displacement equals the charge surface density σ. $\nabla \cdot D = \sigma$.

Unit: Coulomb per metre squared. $L^{-2} T I$; $M^{\frac{1}{2}} L^{-3/2} \mu^{-\frac{1}{2}}$; $M^{\frac{1}{2}} L^{-\frac{1}{2}} T^{-1} \varepsilon^{\frac{1}{2}}$ (vector).
Note: (1) The quantity is also termed the electric flux density.
(2) (U/R) $\overset{*}{D} = 4\pi D$; $\nabla \cdot \overset{*}{D} = 4\pi\sigma$. The unrationalised displacement is also termed the electric induction.

DISSIPATION COEFFICIENT

See scattering coefficient (note 1)

DISTANCE

See length (note)

EFFICIENCY η (O)

Definition: Power output P_o divided by power input P_i. $\eta = P_o/P_i$.
Unit: There are no units or dimensions (scalar).
Note: Efficiency is usually expressed as a percentage.

EFFICIENCY, LUMINOUS

(1) For radiation
   Symbol: K_λ, with K_m for maximum value (photoptic; the corresponding scotoptic symbols are K_λ', K_m').
   Definition: Luminous flux Φ_λ at a given wavelength divided by the corresponding radiant flux P. $K_\lambda = \Phi_\lambda/P$.
   Note: The quantity can be measured for the complete visual wavelength range, and is represented by K (photoptic; scotoptic K').
(2) For a source
   Symbol: η (O).
   Definition: Luminous flux Φ emitted divided by the power P consumed by the source. $\eta = \Phi/P$.
Unit: Lumen per watt. $M^{-1} L^{-2} T^3 F$; no mechanical units (scalar).

EFFICIENCY, LUMINOUS, RELATIVE

Symbol: V_λ (photoptic; the corresponding scotoptic symbol is V_λ').
Definition: Luminous efficiency K_λ divided by maximum luminous efficiency K_m. $V_\lambda = K_\lambda/K_m$. Alternatively, the relative luminous efficiency is the radiant flux at a given wavelength divided by that at the wavelength which produces equally intense luminous sensations under specified photometric conditions, the maximum value of the ratio being made equal to unity.
Unit There are no units or dimensions (scalar).

Note: The internationally accepted values for V_λ are given in the following table, the values of the wavelengths λ being in nanometres.

λ	00	10	20	30	40
300	—	—	—	—	—
400	0·000 4	0·001 2	0·004 0	0·011 6	0·023
500	0·323	0·503	0·710	0·862	0·954
600	0·631	0·503	0·381	0·265	0·175
700	0·004 1	0·002 1	0·001 05	0·000 52	0·000 25

λ	50	60	70	80	90
300	—	—	—	0·000 04	0·000 12
400	0·038	0·060	0·091	0·139	0·208
500	0·995	0·995	0·952	0·870	0·757
600	0·107	0·061	0·032	0·017	0·008 2
700	0·000 12	0·000 06	0·000 03	0·000 015	—

EFFORT

See mechanical advantage

ELASTANCE

Definition: The reciprocal of the capacitance C. Elastance $= 1/C$.
Unit: Per farad (daraf). $M L^2 T^{-4} I^{-2}$; $L T^{-2} \mu$; $L^{-1} \varepsilon^{-1}$ (scalar).

ELASTIC MODULUS

(1) YOUNG'S MODULUS E
   Definition: Normal stress σ divided by tensile strain e. $E = \sigma/e$.
(2) RIGIDITY MODULUS G
   Definition: Shear stress τ divided by shear strain γ. $G = \tau/\gamma$.
   Note: The quantity is also termed the shear modulus.
(3) BULK MODULUS K
   Definition: Bulk stress p divided by bulk strain θ. $K = p/\theta$.
   Note: The quantity is also termed the hydrostatic modulus or modulus
   of compression.
Unit: Pascal. $M L^{-1} T^{-2}$ (scalar).

ELECTRISATION

Symbol: E_i; sometimes K_i.
Definition: Electric polarisation P divided by the electric constant ε_0.
$E_i = P/\varepsilon_0 = \overset{*}{P}/\overset{*}{\varepsilon_0}$.
Unit: Volt per metre = newton per coulomb. $M\ L\ T^{-3}\ I^{-1}$; $M^{\frac{1}{2}}\ L^{\frac{1}{2}}\ T^{-2}\ \mu^{\frac{1}{2}}$; $M^{\frac{1}{2}}\ L^{-\frac{1}{2}}\ T^{-1}\ \varepsilon^{-\frac{1}{2}}$ (vector).

ELECTROMOTIVE FORCE E

Definition: Work done in taking a charge completely round a circuit, ie the circular integral of the electric field strength K. $E = \oint K\ dx$ (x = distance).
Unit: Volt. $M\ L^2\ T^{-3}\ I^{-1}$; $M^{\frac{1}{2}}\ L^{3/2}\ T^{-2}\ \mu^{\frac{1}{2}}$; $M^{\frac{1}{2}}\ L^{\frac{1}{2}}\ T^{-1}\ \varepsilon^{-\frac{1}{2}}$ (scalar).

ELONGATION, FRACTIONAL (or RELATIVE)

See strain (1) (note)

EMITTANCE, LUMINOUS

Symbol: M; occasionally M_v.
Definition: Luminous flux Φ emergent from an infinitesimal element of surface containing the point whose luminous emittance is to be determined, divided by the area A of the element. $M = d\Phi/dA$.
Unit: Lux. $L^{-2}\ F$; $M\ T^{-3}$ (scalar).
Note: The quantity is also termed the luminous exitance.

EMITTANCE, RADIANT

Symbol: M; sometimes M_e.
Definition: Radiant flux Φ emergent from an infinitesimal element of surface containing the point whose radiant emittance is to be determined, divided by the area A of the element. $M = d\Phi/dA$.
Unit: Watt per metre squared. $L^{-2}\ T^{-1}\ Q$; $M\ T^{-3}$ (scalar).
Note: The quantity is also termed the radiant exitance, and is proportional in size to the fourth power of the absolute temperature T of the emitter in the case of a black body (Stefan's law): $M = \sigma T^4$, where σ is Stefan's constant. It has also been termed the radiant flux density and the radiancy.

ENERGY E

Definition: The product of force F and the distance s moved in the direction of the force. $E = F\ s$.
Unit: Joule. $M\ L^2\ T^{-2}$ (scalar).

Note: There are several special types of energy:

enthalpy	H; sometimes I	also called heat function: $H = U + pV$
exergy	E	
free energy	F	also called Helmholtz function: $F = U - TS$
Gibbs function	G	also called free enthalpy: $G = U + pV - TS$
heat (energy) = quantity of heat	Q	formerly also called thermal charge, measured in thermal coulombs (see entropy, note)
internal energy	U; sometimes E	
latent heat (of transformation)	L	quantity of heat energy absorbed or released in an isothermal change of phase
kinetic energy	T; sometimes E_k or K	$T = \frac{1}{2}mv^2$
potential energy	U; sometimes V, E_p or Φ	$U = mgh$
radiant energy	Q; sometimes W, Q_e or U	$Q = \int \Phi_e dt$
work	W; sometimes A	mechanical energy

g = acceleration of free fall S = entropy v = velocity
h = height above a datum level t = time V = volume
m = mass T = absolute temperature Φ_e = radiant flux
p = pressure

ENERGY DENSITY

Symbol: w; E in acoustics; u for radiant energy density.
Definition: Energy E divided by volume V at a point. $w = E/V$.
Unit: Joule per metre cubed. $M\ L^{-1}\ T^{-2}$ (scalar).

ENERGY, LUMINOUS

Symbol: Q; occasionally Q_v.
Definition: The integral of the luminous flux Φ with respect to time t.
$Q = \int \Phi\ dt$.
Unit: Talbot (or lumberg). $T\ F$; $M\ L^2\ T^{-2}$ (scalar).

ENERGY, MOLAR

Definition: Energy E divided by molar value n. Molar energy $= E/n$.
Unit: Joule per mole. $Q\ N^{-1}$; $M\ L^2\ T^{-2}\ N^{-1}$ (scalar).
Note: N is used as the dimensional symbol of molar value.

ENERGY, SPECIFIC   *e*

Definition: Energy E divided by mass m. $e = E/m$.
Unit: Joule per kilogramme. $L^2 T^{-2}$ (scalar).
Note: All the types of energy given under energy (note) have specific forms, defined as above; their symbols are the lowercase equivalents of the symbols tabulated.

ENERGY, SPECTRAL RADIANT U_λ (O)

Definition: Rate of variation of radiant energy U with wavelength λ.
$U_\lambda = dU/d\lambda$.
Unit: Joule per metre. $L^{-1} Q$; $M L T^{-2}$ (scalar).

ENTHALPY

See energy (note)

ENTROPY

Symbol: S; formerly Φ.
Definition: When a small quantity of heat energy dQ is received by a system in which no irreversible change occurs and at a thermodynamic temperature T, the entropy of the system increases by dQ/T. $\Delta S = dQ/T$.
Unit: Joule per kelvin. $Q \ominus^{-1}$; M (scalar).
Note: The quantity is also termed the thermal charge; the unit is then termed the thermal coulomb.

ENTROPY, MOLAR

Definition: Entropy S divided by molar value n. Molar entropy $= S/n$.
Unit: Joule per mole kelvin. $Q \ominus^{-1} N^{-1}$; $M N^{-1}$ (scalar).
Note: N is used as the dimensional symbol of molar value.

ENTROPY, SPECIFIC   *s*

Definition: Entropy S divided by mass m. $s = S/m$.
Unit: Joule per kilogramme kelvin. $M^{-1} Q \ominus^{-1}$; no mechanical units (scalar).

EXERGY

See energy (note)

EXITANCE, LUMINOUS

See emittance, luminous (note)

EXITANCE, RADIANT

See emittance, radiant (note)

EXPANSION COEFFICIENT

(1) LINEAR EXPANSION COEFFICIENT
  Symbol: α; occasionally λ or α_t.
  Definition: Fractional increase in length l divided by increase in tempera-
  ture θ under specified conditions. $\alpha = \dfrac{\Delta l}{l\,\Delta\theta} \approx \dfrac{1}{l}\dfrac{dl}{d\theta}$.

(2) AREAL EXPANSION COEFFICIENT β (O)
  Definition: Fractional increase in area A divided by increase in tempera-
  ture θ under specified conditions. $\beta = \dfrac{\Delta A}{A\,\Delta\theta} \approx \dfrac{1}{A}\dfrac{dA}{d\theta}$.

  Note: The quantity is also termed the superficial expansion coefficient.

(3) CUBICAL EXPANSION COEFFICIENT
  Symbol: γ; occasionally α or α_V or β.
  Definition: Fractional increase in volume V divided by increase in
  temperature θ under specified conditions. $\gamma = \dfrac{\Delta V}{V\,\Delta\theta} \approx \dfrac{1}{V}\dfrac{dV}{d\theta}$.

  Note: The quantity is also termed the volume expansion coefficient or
  the expansivity.
Unit: Per kelvin. Θ^{-1}; $L^{-2}\,T^2$ (scalar).
Note: The condition specified is generally standard pressure. Because the
expansion coefficient varies with temperature, the quantity is usually quoted
at a given temperature (eg 15 °C) or as a mean value over a given temperature
range (eg 0 °C to 100 °C).

EXPANSIVITY

See expansion coefficient (3) (note)

EXTINCTION COEFFICIENT *e* (O)

Definition: The natural logarithm of the reciprocal of the transmissivity t.
$e = \ln(1/t)$.
Unit: There are no units or dimensions (scalar).

FIELD STRENGTH, ELECTRIC

Symbol: E. K is also used, especially when E is required for electromotive
force.
Definition: Force F exerted by an electric field on an electric charge Q
divided by that charge. $E = F/Q$.

Unit: Volt per metre = newton per coulomb. M L T^{-3} I^{-1}; M$^{\frac{1}{2}}$ L$^{\frac{1}{2}}$ T^{-2} $\mu^{\frac{1}{2}}$;
M$^{\frac{1}{2}}$ L$^{-\frac{1}{2}}$ T^{-1} $\varepsilon^{-\frac{1}{2}}$ (vector).
Note: The quantity is also termed the electric field intensity.

FIELD STRENGTH, GRAVITATIONAL R (O)

Definition: The force F exerted by a gravitational field on a mass m divided
by that mass. $R = F/m$.
Unit: Newton per kilogramme. L T^{-2} (vector).

FIELD STRENGTH, MAGNETIC H

Definition: The curl of the magnetic field strength is equal to the sum of the
current density J and the rate of change of displacement D with respect to
time t (Maxwell's equation). $\nabla \times H = J + \dfrac{\partial D}{\partial t}$.

Unit: Ampère per metre. L^{-1} I; M$^{\frac{1}{2}}$ L$^{-\frac{1}{2}}$ T^{-1} $\mu^{-\frac{1}{2}}$; M$^{\frac{1}{2}}$ L$^{\frac{1}{2}}$ T^{-2} $\varepsilon^{\frac{1}{2}}$ (vector).
Note: (1) When referring to the magnetic field strength of the earth's field
the symbols used are H_0 for the horizontal component, V for the vertical
component, and R for the resultant.

(2) (U/R) $\overset{\bullet}{H} = 4\pi H$; $\nabla \times \overset{\bullet}{H} = 4\pi J + \dfrac{\partial \overset{\bullet}{D}}{\partial t}$.

FLUIDITY ϕ

Definition: The reciprocal of the (dynamic) viscosity η. $\phi = 1/\eta$.
Unit: Metre squared per newton second. M^{-1} L T (scalar).

FLUIDITY, THERMAL

See resistivity, thermal (note)

FLUX DENSITY, ELECTRIC

See displacement, electric (note 1)

FLUX DENSITY, MAGNETIC B

Definition: The vector product of the magnetic flux density and the current I
is equal to the force F per unit length s. $B \times I = F/s$.
Unit: Tesla. M T^{-2} I^{-1}; M$^{\frac{1}{2}}$ L$^{-\frac{1}{2}}$ T^{-1} $\mu^{\frac{1}{2}}$; M$^{\frac{1}{2}}$ L$^{-3/2}$ $\varepsilon^{-\frac{1}{2}}$ (vector).
Note: The quantity is also termed the magnetic induction.

FLUX DENSITY, MAGNETIC INTRINSIC

See polarisation, magnetic (note 1)

FLUX DENSITY, RADIANT

See emittance, radiant (note)

FLUX, ENERGY

See power (note 1)

FLUX, LUMINOUS

Symbol: Φ; sometimes F (BS) or Φ_v.
Definition: That quantity characteristic of radiant flux which expresses its capacity to produce a visual sensation, evaluated according to the values of the relative luminous efficiency for the light-adapted eye adopted by the Commission Internationale de l'Eclairage. It is the rate of flow of luminous energy, ie the luminous power.
Unit: Lumen. F; $M L^2 T^{-3}$ (scalar).
Note: The luminous flux emitted by a source is equal to the mean luminous intensity of the source in all directions in space, multiplied by 4π.

FLUX, MAGNETIC Φ

Definition: The scalar product of the magnetic flux density B and the area A.
$\Phi = BA$.
Unit: Weber. $M L^2 T^{-2} I^{-1}$; $M^{\frac{1}{2}} L^{3/2} T^{-1} \mu^{\frac{1}{2}}$; $M^{\frac{1}{2}} L^{\frac{1}{2}} \varepsilon^{-\frac{1}{2}}$ (scalar).

FLUX (OF DISPLACEMENT), ELECTRIC Ψ

Definition: The scalar product of the displacement D and the area A.
$\Psi = DA$.
Unit: Coulomb. $T I$; $M^{\frac{1}{2}} L^{\frac{1}{2}} \mu^{-\frac{1}{2}}$; $M^{\frac{1}{2}} L^{3/2} T^{-1} \varepsilon^{\frac{1}{2}}$ (scalar).
Note: (U/R) $\overset{*}{\Psi} = 4\pi\Psi$; $\overset{*}{\Psi} = \overset{*}{D}A$.

FLUX, RADIANT

Symbol: Φ; sometimes P or Φ_e.
Definition: Rate of flow of radiant energy U with respect to time t, ie radiant power. $\Phi = dU/dt$.
Unit: Watt. $T^{-1} Q$; $M L^2 T^{-3}$ (scalar).
Note: The quantity has also been termed the radiance.

FLUX, SOUND

Symbol: P (BS) or Φ; sometimes N or W.
Definition: The mean rate of flow of sound energy E with respect to time t through an area normal to the direction of flow, ie acoustical power.

$P = dE/dt$. For a plane or spherical progressive wave with a velocity of propagation c in an isotropic medium of density ρ, the sound flux through an area A at a point where the root-mean-square sound pressure is p reduces to $P = p^2 \ A/\rho c$.
Unit: Watt. M L² T⁻³ (scalar).

FORCE

Symbol: F; sometimes P.
Definition: Rate of change of momentum p with respect to time t. $F = dp/dt$. When the mass m is constant, force is the product of mass and acceleration a. $F = m \ a$.
Unit: Newton. M L T⁻² (vector).
Note: (1) The quantity is also termed tension or thrust.
(2) Weight (W; occasionally G or P) is the vertical force acting on a body as a result of the existence of gravity; it is that force which, when applied to the body, gives it an acceleration equal to the local acceleration of free fall. The local acceleration of free fall includes both the gravitational and centrifugal components, but excludes the effect of atmospheric buoyancy.

FREQUENCY

Symbol: f. the symbol v, formerly widely used, is no longer preferred.
Definition: The reciprocal of periodic time T. $f = 1/T$.
Unit: Hertz. T⁻¹ (scalar).
Note: The term pitch is used as a measure of the musical awareness of frequency. Eg the pitch of the note A in the treble stave is a frequency of 440 hertz.

FREQUENCY, ANGULAR (or CIRCULAR)

See velocity, angular (note 1)

FREQUENCY, ROTATIONAL n

Definition: Number of revolutions N divided by time t. $n = N/t$.
Unit: Hertz. T⁻¹ (scalar).
Note: This quantity, also termed rotational speed, must be distinguished from angular velocity.

FRICTION COEFFICIENT

(1) STATIC FRICTION COEFFICIENT
    Symbol: μ; rarely f.
    Definition: Limiting statical frictional force F divided by normal reaction R. $\mu = F/R$.

(2) DYNAMIC FRICTION COEFFICIENT μ' (O)

Definition: Limiting dynamical frictional force F' divided by normal reaction R . $\mu' = F'/R$.
Note: The quantity is also termed kinematic friction coefficient.
Unit: There are no units or dimensions (scalar).
Note: The quantity is also called factor of friction. The value of the static coefficient between two surfaces exceeds the value of the dynamic coefficient between the same two surfaces.

FUEL CONSUMPTION

See traffic factor (1)

GIBBS FUNCTION

See energy (note)

GRAVITY, SPECIFIC

See density, relative (note 1)

HARDNESS

There are a number of arbitrarily defined meanings of this quantity. See the entries in Part 1 'degree (of hardness)' and 'hardness number', and Table 55 in Appendix 6.

HEAT CAPACITY

Symbol: c ; sometimes C (BS).
Definition: The rate of increase of heat energy Q with temperature θ under specified conditions. $c = dQ/d\theta$.
Unit: Joule per kelvin. $Q \Theta^{-1}$; M (scalar).

HEAT FLOW RATE

See power (note 2)

HEAT FLOW RATE DENSITY

Symbol: q ; sometimes ϕ .
Definition: Heat flow rate Φ divided by area A . $q = \Phi/A$.
Unit: Watt per metre squared. $L^{-2} T^{-1} Q$; $M T^{-3}$ (scalar).
Note: The quantity is also called heat flow rate intensity.

HEAT, LATENT

See energy (note); latent heat, specific (note)

HEAT, QUANTITY OF

See energy (note)

HEAT RELEASE H (O)

Definition: Quantity of heat Q released (in furnaces, etc) divided by the product of the volume V and time t. $H = Q/Vt$.
Unit: Watt per metre cubed. $L^{-3} T^{-1} Q$; $M L^{-1} T^{-3}$ (scalar).

HUMIDITY, ABSOLUTE d_v (BS)

Definition: The mass m of water vapour present in a sample of moist air divided by the volume V of moist air. $d_v = m/V$.
Unit: Kilogramme per metre cubed. $M L^{-3}$ (scalar).
Note: The quantity is also termed the vapour concentration.

HUMIDITY, RELATIVE

Symbol: ϕ_p; sometimes U (BS).
Definition: Actual vapour pressure p divided by the saturation vapour pressure p_0 over a plane liquid-water surface at the same (dry-bulb) temperature. $\phi_p = p/p_0$.
Unit: There are no units or dimensions (scalar).
Note: Relative humidity is usually expressed as a percentage.

HUMIDITY, SPECIFIC

Symbol: x; sometimes q (BS).
Definition: Mass m of water vapour present in a sample of moist air divided by the mass m_0 of moist air. $x = m/m_0$.
Unit: There are no units or dimensions (scalar).

HYDROSTATIC MODULUS

See elastic modulus (3)

ILUMINANCE

See illumination (note)

ILLUMINATION (INTENSITY OF)

Symbol: E; occasionally E_v.
Definition: The illumination at a point surface is the luminous flux Φ incident on an infinitesimal element of surface containing the point divided by the area A of the element. $E = d\Phi/dA$.
Unit: Lux. L^{-2} F; M T^{-3} (scalar).
Note: The quantity is also termed the illuminance.

IMPEDANCE, ACOUSTICAL

Symbol: Z_a; occasionally Z.
Definition: Complex ratio of the sound pressure p to the strength U of the sound. $Z_a = p/U$.
Unit: Pascal second per metre cubed. M L^{-4} T^{-1} (vector).
Note: (1) The acoustical impedance refers to a surface vibrating so as to constitute a simple sinusoidal source of sound.
(2) $Z_a = R_a + i\ X_a$, where R_a = acoustical impedance and X_a = acoustical reactance.

IMPEDANCE, ELECTRICAL Z

Definition: Complex representation of potential difference V divided by the complex representation of current I. $Z = V/I$.
Unit: Ohm. M L^2 T^{-3} I^{-2}; $L\ T^{-1}\ \mu$; $L^{-1}\ T\ \varepsilon^{-1}$ (scalar).
Note: $Z = R + i\ X$ (R = resistance, X = reactance). $Z = |Z|\ e^{i\varphi}$ (ϕ = phase displacement); hence $|Z| = (R^2 + X^2)^{\frac{1}{2}}$. The term impedance is sometimes given to the modulus $|Z|$, Z being termed the complex impedance.

IMPEDANCE, MECHANICAL

Symbol: Z_m; occasionally w.
Definition: Complex ratio of the force F acting in the direction of motion to the velocity v at a point or surface. $Z_m = F/v$.
Unit: Newton second per metre. M T^{-1} (vector).
Note: $Z_m = R_m + i\ X_m$, where R_m = mechanical resistance and X_m = mechanical reactance.

IMPEDANCE, SPECIFIC ACOUSTICAL

Symbol: Z_s; sometimes W.
Definition: Complex ratio of the sound pressure p to the sound particle velocity v. $Z_s = p/v$. Alternatively, the specific acoustical impedance is the product of the acoustical impedance Z_a and the area A. $Z_s = Z_a A$.
Unit: Pascal second per metre. M L^{-2} T^{-1} (vector).

Note: (1) The quantity was formerly called the unit-area acoustical impedance.

(2) The specific acoustical impedance refers to a point in a medium in which sound waves are propagated.

(3) $Z_s = R_s + i X_s$, where R_s = specific acoustical resistance and X_s = specific acoustical reactance.

IMPULSE I (O)

Definition: The integral of force F with respect to time t. $I = \int F \, dt$.
Unit: Newton second. M L T^{-1} (vector).

INDUCTANCE

(1) SELF INDUCTANCE L
   Definition: Magnetic flux Φ through a loop caused by a current I in the loop, divided by the current. $L = \Phi/I$.
(2) MUTUAL INDUCTANCE
   Symbol: M or L_{12}
   Definition: Magnetic flux Φ through a loop caused by a current I in another loop, divided by the current. $M = \Phi/I$.
Unit: Henry. M L^2 T^{-2} I^{-2}; L μ; L^{-1} T^2 ε^{-1} (scalar).

INDUCTANCE, THERMAL

Definition: The product of the temperature difference θ and time t, divided by the mean rate of flow of entropy S with respect to time.

$$\text{Thermal inductance} = \theta \, t \left/ \frac{dS}{dt} \right.$$

Unit: Thermal henry = kelvin squared second per watt. T^2 Q^{-1} Θ^2; M^{-1} L^2 (scalar).

Note: The former definition was: the product of the temperature difference and the time divided by the rate of flow of heat energy Q with respect to time.

$$\text{Thermal inductance} = \theta \, t \left/ \frac{dQ}{dt} \right.$$

INDUCTION, ELECTRIC

See displacement, electric (note 2)

INDUCTION, MAGNETIC

See flux density, magnetic (note)

INTENSITY LEVEL N (O)

Definition: The common logarithm of the ratio of two intensities I_1, I_2 or two powers P_1, P_2 or two energies E_1, E_2, each expressed in the same units. $N = \log (I_2/I_1) = \log (P_2/P_1) = \log (E_2/E_1)$.

Unit: Bel; in practice the decibel is always used. There are no dimensions (scalar).

Note: (1) The symbols L_P, L_N and L_W are also used for power level.

(2) In acoustics the conventional reference levels are the rounded values of the threshold level of aural sensation; $I_1 = 10^{-12}$ watts per metre squared, $P_1 = 10^{-12}$ watts, $E_1 = 10^{-12}$ joules.

(3) Intensity level is also equivalent to (sound) pressure level (L_p or L), defined as twice the common logarithm of the ratio of two root-mean-square pressures p_1, p_2, each expressed in the same units. $N = 2 \log (p_2/p_1)$. The conventional reference levels are $p_1 = 2 \times 10^{-5}$ pascals in air and $p_1 = 0 \cdot 1$ pascal in water.

(4) See the entry sensation level.

INTENSITY, LUMINOUS

Symbol: I; occasionally I_v.

Definition: The luminous intensity in a given direction is the luminous flux Φ emitted by a point source in an infinitesimal cone containing the direction divided by the solid angle ω of the cone. $I = d\Phi/d\omega$.

Unit: Candela. F; M L^2 T^{-3} (scalar).

Note: (1) The quantity has also been given the deprecated name candle power.

(2) The standard source is a full (ie planckian) radiator at the temperature of the solidification of platinum (2 045 K).

(3) Mean spherical intensity is the average value of the luminous intensity of a source in all directions.

(4) The luminous intensity of a source can be compared to that of a standard tungsten filament lamp to an accuracy (at the National Physical Laboratory) of one part in 5 000.

INTENSITY, RADIANT

Symbol: I; sometimes I_e.

Definition: The radiant intensity in a given direction is the radiant flux Φ emitted by a point source in an infinitesimal cone containing the direction divided by the solid angle ω of the cone. $I = d\Phi/d\omega$.

Unit: Watt per steradian. T^{-1} Q; M L^2 T^{-3} (scalar).

INTENSITY, SOUND

Symbol: I; occasionally J.

Definition: Mean rate of flow of sound energy E per unit area A normal to the direction of propagation of the sound wave. $I = \dfrac{1}{A}\dfrac{dE}{dt}$ (t = time). For a plane or spherical progressive wave with a velocity of propagation c in an isotropic medium of density ρ, the sound intensity at a point where the root-mean-square sound pressure is p reduces to $I = p^2/\rho c$.

Unit: Watt per metre squared. M T^{-3} (scalar).

INTERVAL I (O)

Definition: The relation between the two pitches measured either as the ratio of the frequencies f_1, f_2 of corresponding pure tones (ie standard sinusoidal plane progressive tones of equal pitches), or in terms of the logarithm of their ratio. The ratio is generally expressed as a fraction greater than unity. Either $I = f_2/f_1$ or $I = k \log (f_2/f_1)$.

Unit: For I as a ratio, there are no units. For I expressed in terms of the logarithm of a ratio, the units are the octave ($k = 1/\log 2$), the cent ($k = 1\,200/\log 2$), the savart ($k = 1\,000$) and the modified savart ($k = 300/\log 2$). There are no dimensions (scalar).

Note: (1) The quantity is also termed the pitch (or frequency) interval.

(2) The term musical interval refers to a pitch interval of particular importance defined according to one of a number of rules derived from a combination of empirical considerations and mathematical calculations. The sizes of the more important intervals are given in table 57 of Appendix 7.

IRRADIANCE

Symbol: E; sometimes E_e.

Definition: The irradiance at a point surface is the radiant flux Φ incident on an infinitesimal element of surface containing the point divided by the area A of the element. $E = d\Phi/dA$.

Unit: Watt per metre squared. L^{-2} T^{-1} Q; M T^{-3} (scalar).

Note: The quantity has also been termed the irradiancy.

LATENT HEAT, SPECIFIC l

Definition: Quantity of heat energy Q absorbed or released in an isothermal transformation of phase, divided by mass m. $l = Q/m$.

Unit: Joule per kilogramme. M^{-1} Q; L^2 T^{-2} (scalar).

Note: The quantity is often erroneously referred to by its former name, latent heat.

LEAKAGE COEFFICIENT

See coupling coefficient (note)

LENGTH *l*

Definition: There is no formal definition of length. It can only be explained in terms of an inadequate synonym, eg dimension.
Unit: Metre. L (vector).
Note: The following quantities (with their symbols) are types of length:

altitude	*z* (BS)		
breadth	*b*		
depth	*h*		
diameter	*d*	*D* (O)	
displacement	*s*		
distance	*s*		
height	*h*		
line segment	*s*		
path length	*s*	*x* (BS)	
radial distance	*r*		
radius	*r*		
thickness	*d*	*t* (O)	δ
wavelength	λ		
width	*b*		

LOAD

See mechanical advantage

LOGARITHMIC DECREMENT

Symbol: Λ; δ (BS).
Definition: The product of the periodic time *T* and the damping coefficient δ.
$\Lambda = \delta T$.
Unit: Neper. There are no dimensions (scalar).
Note: The ratio π/Λ is called the Q-factor.

LOUDNESS

Symbol: *S* or *N*.
Definition: An observer's auditory impression of the volume of a sound referred to a sound of a given equivalent loudness *P*. When the loudness is measured as one sone the given equivalent loudness is 40 phons: in general $S = 2^{(P-40)/10}$.
Unit: Sone. There are no dimensions (scalar).

LOUDNESS, EQUIVALENT

Symbol: P or L_N; occasionally Λ.

Definition: A quantity measured by the sound pressure level of a standard pure tone of given frequency coming from directly in front of the observer, subjectively judged to be of equal loudness by an otologically normal ear. (The mean, or rather the modal value, of a series of readings should be taken.) The standard pure tone of given frequency is a sinusoidal plane progressive wave of frequency 1 000 hertz.

Unit: Phon. There are no dimensions (scalar).

Note: The quantity is also termed the loudness level.

LUMINANCE

Symbol: L; occasionally L_V.

Definition: That property by virtue of which a surface emits more or less light in the direction of view. The luminance at a point surface in a given direction is the luminous intensity I in the given direction of an infinitesimal element of surface containing the point divided by the orthogonally projected area A of the element in a plane normal to the given direction. $L = dI/dA$. An alternative definition, used by engineers (but deprecated), is the luminance of the equivalent diffusing surface that emits or reflects light with such a spatial distribution that the surface has the same luminance whatever the direction from which it is viewed. $L = \rho I$ (ρ = reflexion coefficient).

Unit: Nit. L^{-2} F; $M\ T^{-3}$ (scalar).

Note: The quantity has also been given the deprecated name brightness.

LUMINANCE, EQUIVALENT

Definition: The luminance of a light at a colour temperature of a full (ie planckian) radiator at the temperature of solidification of platinum which under specified conditions has a luminosity equal to that of the light considered.

Unit: Nit. L^{-2} F; $M\ T^{-3}$ (scalar).

Note: (1) The quantity equivalent luminance is required for coloured lights at low luminance because of the Purkinje effect: a comparison of the luminosities of differently coloured sources depends on the luminance levels of the sources, particularly below about 10 nit.

(2) The temperature of solidfication of platinum = 2 045 K.

LUMINANCE FACTOR

Definition: The luminance of a body when illuminated and observed under certain conditions divided by that of a perfect diffuser under identical conditions.

Unit: There are no units or dimensions (scalar).
Note: The luminance factor depends on the angle of incidence, mode of illumination and spectral composition of the incident light.

LUMINOSITY

Definition: The attribute of visual perception such that an area appears to emit more or less light.
Unit: Nit. L^{-2} F; M T^{-3} (scalar).
Note: The quantity is also termed the apparent brightness or the subjective brightness.

MAGNETISATION

Symbol: M; occasionally H_i.
Definition: The ratio of the magnetic flux density B to the magnetic constant μ_0, minus the magnetic field strength H. $M = \dfrac{B}{\mu_0} - H$.
Unit: Ampère per metre. L^{-1} I; $M^{\frac{1}{2}} L^{-\frac{1}{2}} T^{-1} \mu^{-\frac{1}{2}}$; $M^{\frac{1}{2}} L^{\frac{1}{2}} T^{-2} \varepsilon^{\frac{1}{2}}$ (vector).
Note: (U/R) $\overset{*}{M} = 4\mu M$; $\overset{*}{M} = \dfrac{B}{\overset{*}{\mu_0}} - \overset{*}{H}$.

MAGNETOMOTIVE FORCE

Symbol: F; sometimes F_m or $\mathscr{F}$.
Definition: The circular integral of the magnetic field strength H. $F = \oint H \, dx$ (x = distance).
Unit: Ampère(-turn). I; $M^{\frac{1}{2}} L^{\frac{1}{2}} T^{-1} \mu^{-\frac{1}{2}}$; $M^{\frac{1}{2}} L^{3/2} T^{-2} \varepsilon^{\frac{1}{2}}$ (scalar).
Note: (U/R) $\overset{*}{F} = 4\pi F$; $\overset{*}{F} = \oint \overset{*}{H} \, dx$.

MASS *m*

Definition: There is no formal definition of mass. It can only be explained in terms of an inadequate synonym, eg quantity of matter.
Unit: Kilogramme. M (scalar).

MASS, ACOUSTICAL m_a (BS)

Definition: Acoustical mass reactance X_a divided by angular frequency ω. $m_a = x_a/\omega$.
Unit: Pascal second squared per metre cubed. M L^{-4} (vector).

MASS DENSITY

See density (note 1)

MASS-DISTANCE

See traffic factor (2)

MASS FLOW RATE q_m

Definition: Mass m divided by time t of flow. $q_m = m/t$.
Unit: Kilogramme per second. $M\ T^{-1}$ (scalar).

MECHANICAL ADVANTAGE M (O)

Definition: The load (output force) F_1 moved by a machine during the application of a certain effort (imput force), divided by the effort F_2. $M = F_1/F_2$.
Unit: There are no units or dimensions (scalar).

MESOPTIC...

All luminous quantities as defined may apply to the mesoptic system of photometry, the range over which the spectral sensitivity of the human eye is changing from the photoptic (qv) to the scotoptic (qv) states. The adjective *mesoptic* precedes the name of the quantity. At present it is not possible to define mesoptic terms exactly, since the mesoptic range has not yet been fully studied.

MIXING RATIO r (BS)

Definition: Mass m of water vapour present in a sample of air divided by the mass m_0 of dry air with which the vapour is associated. $r = m/m_0$.
Unit: There are no units or dimensions (scalar).

MOLALITY m (BS)

Definition: Molar value n of solute divided by mass M of solvent. $m = n/M$.
Unit: Mole per kilogramme. $M^{-1}\ N$ (scalar).
Note: N is used as the dimensional symbol of molar value.

MOLARITY c (BS)

Definition: Molar value n of solute divided by volume V of solvent. $c = n/V$.
Unit: Mole per metre cubed (unit molarity is, in fact, always given in moles per litre). $L^{-3}\ N$ (scalar).
Note: N is used as the dimensional symbol of molar value.

MOLAR VALUE n (BS)

Definition: The number of moles present.
Unit: Mole. A convenient dimensional symbol is N (scalar).
Note: Other terms beginning 'molar . . .' are entered according to the second word of the term.

MOMENT

Symbol: M; sometimes T or N.
Definition: The moment of a force F about a point is the vector product of the radius vector r from the point to any point on the line of action of the force, and the force. (Ie it is the product of the force and its perpendicular distance from the point.) $M = F \times r$.
Unit: Newton metre (never joule). $M\ L^2\ T^{-2}$ (vector).
Note: The quantity is more correctly termed the moment of a force. It is also termed the moment of a couple or, more loosely, couple; also, torque. The bending moment is a particular kind of moment.

MOMENT, ELECTROMAGNETIC m

Definition: The vector product of the electromagnetic moment and the magnetic flux density B is equal to the moment of the force M. $m \times B = M$.
Unit: Ampère metre squared. $L^2\ I$; $M^{\frac{1}{2}}\ L^{5/2}\ T^{-1}\ \mu^{-\frac{1}{2}}$; $M^{\frac{1}{2}}\ L^{7/2}\ T^{-2}\ \varepsilon^{\frac{1}{2}}$ (vector).
Note: The quantity is also termed the magnetic moment or the magnetic area moment; it should not be confused with the magnetic dipole moment.

MOMENT, MAGNETIC

See dipole moment, magnetic (note 1); moment, electromagnetic (note)

MOMENT OF INERTIA (DYNAMIC)

Symbol: I; occasionally J.
Definition: The moment of inertia of a body about an axis is the sum of the products of its mass elements m_i and the squares of their distances r_i from the axis. $I = \Sigma m_i\ r_i^2$.
Unit: Kilogramme metre squared. $M\ L^2$ (scalar).

MOMENT, SECOND

(1) SECOND MOMENT OF AREA
Symbol: I or I_a.
Definition: The second moment of area of a plane area about an axis in its plane is the sum of the products of its elements of area A_i and the squares of their distances r_i from the axis. $I = \Sigma A_i\ r_i^2$.

Note: The quantity is better termed the second axial moment of area, or alternatively the geometrical moment of inertia. It must not be confused with the moment of inertia.

(2) SECOND POLAR MOMENT OF AREA

Symbol: I_p or J.

Definition: The second polar moment of area of a plane area about a point in its plane is the sum of the products of its elements of area A_i and the squares of their distances r_i from the point. $I_p = \Sigma A_i\, r_i^2$.

Unit: Metre to the fourth power. L^4 (scalar).

MOMENTUM, ANGULAR

Symbol: b; sometimes p_θ.

Definition: The angular momentum of a particle about a point is the vector product of the radius vector r from the point to the particle and the momentum p of the particle. $b = r \times p$. Alternatively, angular momentum is the product of moment of inertia I and angular velocity ω. $b = I\omega$.

Unit: Kilogramme metre squared per second. $M\ L^2\ T^{-1}$ (vector).

Note: The quantity is also termed the moment of momentum.

MOMENTUM (TRANSLATIONAL) p

Definition: The product of mass m and velocity v. $p = mv$.

Unit: Kilogramme metre per second. $M\ L\ T^{-1}$ (vector).

NOISE LEVEL, PERCEIVED

Definition: A quantity measured by the sound pressure level of a reference sound subjectively judged by an observer to be equally noisy. The reference sound consists of a band of random noise of width one-third to one octave centred on a frequency of one thousand hertz.

Unit: Perceived noise decibel. There are no dimensions (scalar).

OPACITY

Definition: The reciprocal of the transmission coefficient τ. Opacity $= 1/\tau$.

Unit: There are no units or dimensions (scalar).

PELTIER COEFFICIENT a_P

Definition: Heat energy E liberated or absorbed at a thermoelectric junction divided by the charge Q flowing through the junction. $a_P = E/Q$.

Unit: Joule per coulomb = volt. $M\ L^2\ T^{-3}\ I^{-1}$; $M^{\frac{1}{2}}\ L^{3/2}\ T^{-2}\ \mu^{-\frac{1}{2}}$; $M^{\frac{1}{2}}\ L^{\frac{1}{2}}\ T^{-1}\ \varepsilon^{-\frac{1}{2}}$ (scalar).

PERIOD

See time (note 1)

PERMEABILITY, ABSOLUTE μ

Definition: Magnetic flux density B divided by magnetic field strength H. $\mu = B/H$.
Unit: Henry per metre. M L T^{-2} I^{-2}; μ; L^{-2} T^2 ε^{-1} (scalar).
Note: (1) The symbol μ_0 is used for the magnetic constant, ie the permeability of free space.
(2) (U/R) $\overset{*}{\mu} = \mu/4\pi$; $\overset{*}{\mu} = B/\overset{*}{H}$.

PERMEABILITY, RELATIVE μ_r

Definition: Absolute permeability μ divided by the magnetic constant μ_0.
$\mu_r = \mu/\mu_0 = \overset{*}{\mu}/\overset{*}{\mu}_0$.
Unit: There are no units or dimensions (scalar).

PERMEANCE

Symbol: Λ; occasionally P.
Definition: The reciprocal of reluctance R. $\Lambda = 1/R$.
Unit: Henry. $_*$M L^2 T^{-2} $_*$I^{-2}; L μ; L^{-1} T^2 ε^{-1} (scalar).
Note: (U/R) $\overset{*}{\Lambda} = \Lambda/4\pi$; $\overset{*}{\Lambda} = 1/\overset{*}{R}$.

PERMITTIVITY, ABSOLUTE ε

Definition: Displacement D divided by electric field strength E. $\varepsilon = D/E$.
Unit: Farad per metre. M^{-1} L^{-3} T^4 I^2; L^{-2} T^2 μ^{-1}; ε (scalar).
Note: (1) The symbol ε_0 is used for the electric constant, ie the permittivity of free space.
(2) (U/R) $\overset{*}{\varepsilon} = 4\pi\varepsilon$; $\overset{*}{\varepsilon} = D/\overset{*}{E}$.
(3) The quantity is also occasionally termed the capacitivity.

PERMITTIVITY, RELATIVE ε_r

Definition: Absolute permittivity ε divided by the electric constant ε_0.
$\varepsilon_r = \varepsilon/\varepsilon_0 = \overset{*}{\varepsilon}/\overset{*}{\varepsilon}_0$.
Unit: There are no units or dimensions (scalar).
Note: The quantity has also been termed the dielectric constant or the specific inductive capacitance.

PHASE COEFFICIENT

See attenuation coefficient

PHASE DISPLACEMENT

See angle (note 2)

PHOTOPTIC ...

All luminous quantities as defined may apply to the photoptic system of photometry, ie with the human eye fully light-adapted. The adjective *photoptic* precedes the name of the quantity. See especially *efficiency, luminous relative.*

PITCH

See frequency (note)

POISSON'S RATIO

Symbol: μ or v.
Definition: Lateral contraction (ie fractional decrease in diameter Δd from the diameter d_o in a specified reference state) divided by the tensile strain e.
$\mu = \Delta d / d_o\, e$.
Unit: There are no units or dimensions (scalar).
Note: The quantity is also termed Poisson's number; it was originally defined (by Poisson) as the reciprocal of the present definition.

POLARISATION, ELECTRIC

Symbol: P; occasionally D_i.
Definition: Displacement D minus the product of the electric constant ε_o and the electric field strength E. $P = D - \varepsilon_o E$.
Unit: Coulomb per metre squared. $L^{-2} T I$; $M^{\frac{1}{2}} L^{-3/2} \mu^{-\frac{1}{2}}$; $M^{\frac{1}{2}} L^{-\frac{1}{2}} T^{-1} \varepsilon^{\frac{1}{2}}$ (vector).
Note: (U/R) $\overset{*}{P} = 4\pi P$; $\overset{*}{P} = \overset{*}{D} - \varepsilon_o \overset{*}{E}$.

POLARISATION, MAGNETIC

Symbol: J; occasionally B_i.
Definition: The product of the magnetisation M and the electric constant μ_o.
$J = \mu_o M$.
Unit: Tesla. $M T^{-2} I^{-1}$; $M^{\frac{1}{2}} L^{-\frac{1}{2}} T^{-1} \mu^{\frac{1}{2}}$; $M^{\frac{1}{2}} L^{-3/2} \varepsilon^{-\frac{1}{2}}$ (vector).
Note: (1) The quantity is also termed the intrinsic magnetic flux density.
(2) (U/R) $\overset{*}{J} = J/4\pi$; $\overset{*}{J} = \overset{*}{\mu_o}\overset{*}{M}/4\pi$.

POLE STRENGTH, MAGNETIC m (O)

Definition: Force F divided by magnetic field strength H. $m = F/H$.
Unit: Newton metre per ampère. $M L^2 T^{-2} I^{-1}$; $M^{\frac{1}{2}} L^{3/2} T^{-1} \mu^{\frac{1}{2}}$; $M^{\frac{1}{2}} L^{\frac{1}{2}} \varepsilon^{-\frac{1}{2}}$
(scalar).
Note: (1) (U/R) $\overset{*}{m} = m/4\pi$; $\overset{*}{m} = F/\overset{*}{H}$.
(2) This quantity is now regarded as associated with an imaginary concept, and is no longer employed.

POTENTIAL DIFFERENCE, ELECTRIC

Symbol: V; sometimes U.
Definition: The potential difference between two points 1, 2 in an electric field of strength E is the line integral from 1 to 2 of E.

$$V = \int_{E_1}^{E_2} E \, dx \ (x = \text{distance}).$$

Unit: Volt. $M L^2 T^{-3} I^{-1}$; $M^{\frac{1}{2}} L^{3/2} T^{-2} \mu^{\frac{1}{2}}$; $M^{\frac{1}{2}} L^{\frac{1}{2}} T^{-1} \varepsilon^{-\frac{1}{2}}$ (scalar).
Note: The quantity is also termed the voltage or the electric tension.

POTENTIAL DIFFERENCE, MAGNETIC

Symbol: U; sometimes U_m or $\mathcal{U}$.
Definition: The potential difference between two points 1, 2 in a magnetic field of strength H is the line integral from 1 to 2 of H.

$$U = \int_{H_1}^{H_2} H \, dx \ (x = \text{distance}).$$

Unit: Ampère(-turn). I; $M^{\frac{1}{2}} L^{\frac{1}{2}} T^{-1} \mu^{-\frac{1}{2}}$; $M^{\frac{1}{2}} L^{3/2} T^{-2} \varepsilon^{\frac{1}{2}}$ (scalar).

Note: (U/R) $\overset{*}{U} = 4\pi U$; $\overset{*}{U} = \int_{\overset{*}{H_1}}^{\overset{*}{H_2}} \overset{*}{H} \, dx$.

POTENTIAL DIFFERENCE, THERMAL

See temperature (3) (note)

POTENTIAL, ELECTRIC

Symbol: V; occasionally ϕ.
Definition: With a reversed sign, the electric field strength E is given by the rate of change of electric potential with respect to distance x. $E = -dV/dx$.
Alternatively, the electric potential is the work W done against an electric field in bringing a charge Q from infinity to its position in the field. $V = W/Q$.
Unit: Volt. $M L^2 T^{-3} I^{-1}$; $M^{\frac{1}{2}} L^{3/2} T^{-2} \mu^{\frac{1}{2}}$; $M^{\frac{1}{2}} L^{\frac{1}{2}} T^{-1} \varepsilon^{-\frac{1}{2}}$ (scalar).

POTENTIAL, GRAVITATIONAL Ω (O)

Definition: With a reversed sign, the gravitational field strength R is given by the rate of change of gravitational potential with respect to distance x. $R = -d\Omega/dx$. Alternatively, the gravitational potential is the work W done by a gravitational field to bring a mass m from infinity to its position in the field. $\Omega = W/m$.
Unit: Joule per kilogramme. $L^2 T^{-2}$ (scalar).

POTENTIAL, MAGNETIC U

Definition: With a reversed sign, the magnetic field strength H is given by the rate of change of magnetic potential with respect to distance x.
$H = -dU/dx$.
Unit: Ampère(-turn). I; $M^{\frac{1}{2}} L^{\frac{1}{2}} T^{-1} \mu^{-\frac{1}{2}}$; $M^{\frac{1}{2}} L^{3/2} T^{-2} \varepsilon^{\frac{1}{2}}$ (scalar).
Note: (U/R) $\overset{*}{U} = 4\pi U$; $\overset{*}{H} = -d\overset{*}{U}/dx$.

POWER P

Definition: Energy E transferred in a certain time t divided by the time.
$P = E/t$.
Unit: Watt. $M L^2 T^{-3}$ (scalar).
Note: (1) The quantity is also termed (energy) flux; it has also been termed activity.
(2) The thermal equivalent is also termed heat flow rate (Φ or q). The radiation equivalent is also termed (radiant) flux: see flux, radiant. The photometric equivalent is also termed (luminous) flux: see flux, luminous. The acoustical equivalent is also termed (sound) flux: see flux, sound.
(3) In alternating current theory, where the potential difference
$$V = \hat{V} \cos \omega t$$
and the current $\qquad\qquad I = \hat{I} \cos (\omega t - \phi)$
(ω = a constant; t = time; ϕ = phase displacement; $\hat{V}, \hat{I}$ = maximum values of V, I), the following types of power occur:

type	symbol	size	unit
active power	P	$I_e V_e \cos \phi$	watt
apparent power	S or P_s	$I_e V_e$	voltampère
reactive power	Q or P_q	$I_e V_e \sin \phi$	var

I_e, V_e = effective values of I, V; $S^2 = P^2 + Q^2$

The ratio P/S is termed the power factor.

POWER LEVEL

See intensity level (note 1)

POWER, THERMOELECTRIC P (O)

Definition: Electromotive force E developed in a thermocouple divided by the difference in temperature ΔT between the junctions. $P = E/\Delta T$.
Unit: Volt per kelvin. $M\ L^2\ T^{-3}\ I^{-1}\ \Theta^{-1}$; $M^{\frac{1}{2}}\ L^{3/2}\ T^{-2}\ \mu^{\frac{1}{2}}\ \Theta^{-1}$; $M^{\frac{1}{2}}\ L^{\frac{1}{2}}\ T^{-1}\ \varepsilon^{-\frac{1}{2}}\ \Theta^{-1}$ (scalar).

POYNTING VECTOR S

Definition: Flux density of electromagnetic field energy, ie the vector product of the electric field strength E and the magnetic field strength H. $S = E \times H$.
Unit: Watt per metre squared. $M\ T^{-3}$ (vector).
Note: If the unrationalised magnetic field strength $\overset{\bullet}{H}$ is used,

$$S = (E \times \overset{\bullet}{H})/4\pi.$$

PRESSURE p

Definition: Force F divided by the area A over which the force acts normally. $p = F/A$.
Unit: Pascal. $M\ L^{-1}\ T^{-2}$ (scalar).
Note: The quantity is also termed the stress or, more properly, normal stress; it is then often represented by σ (or f). In cases where twisting takes place the quantity involved is the shear stress τ (or q). In the case of fluids it is often termed the hydrostatic stress (or pressure) or bulk stress. The somewhat inappropriate term mechanical tension has also been used.

PRESSURE COEFFICIENT β

Definition: Fractional increase in pressure p of a gas divided by increase in temperature θ under specified conditions. $\beta = \dfrac{\Delta p}{p\ \Delta\theta} \approx -\dfrac{1}{p}\dfrac{dp}{d\theta}$.
Unit: Per kelvin. Θ^{-1}; $L^{-2}\ T^2$ (scalar).
Note: The condition specified is generally constant volume.

PRESSURE LEVEL, SOUND

See intensity level (note 3)

PRINCIPAL SPECIFIC HEAT CAPACITIES RATIO

Symbol: γ; occasionally κ.
Definition: Specific heat capacity at constant pressure c_p divided by specific heat capacity at constant volume c_v. $\gamma = c_p/c_v$.
Unit: There are no units or dimensions (scalar).

PROPAGATION COEFFICIENT

See attenuation coefficient

PULSITANCE

See velocity, angular (note 1)

Q-FACTOR

See logarithmic decrement (note)

QUALITY FACTOR Q

Definition: The modulus of the electrical reactance X divided by the electrical resistance R. $Q = |X|/R$.
Unit: There are no units or dimensions (scalar).

RADIANCE

Symbol: L; sometimes L_e.
Definition: The radiance at a point surface in a given direction is the radiant intensity I in the given direction of an infinitesimal element of surface containing the point divided by the orthogonally projected area A of the element in a plane normal to the given direction. $L = dI/dA$.
Unit: Watt per metre squared steradian. $L^{-2} T^{-1} Q$; $M T^{-3}$ (scalar).
Note: The quantity was formerly termed the steradiancy.

RADIANCY

See emittance, radiant (note)

REACTANCE, ACOUSTICAL X_a (BS)

Definition: The magnitude of the imaginary part of the acoustical impedance Z_a. $Z_a = R_a + i X_a$ (R_a = acoustical resistance).
Unit: Pascal second per metre cubed. $M L^{-4} T^{-1}$ (vector).
Note: If the reactance is solely due to inertia it is termed the acoustical mass reactance; if it is solely due to stiffness it is termed the acoustical stiffness reactance.

REACTANCE, ELECTRICAL X

Definition: The magnitude of the imaginary part of the impedance Z.
$Z = R + i X$ (R = resistance).
Unit: Ohm. $M L^2 T^{-3} I^{-2}$; $L T^{-1} \mu$; $L^{-1} T \varepsilon^{-1}$ (scalar).
Note: Reactance due to pure inductance $X_L = \omega L$ ($\omega = 2\pi \times$ frequency, L = inductance); reactance due to pure capacitance $X_C = 1/\omega C$ (C = capacitance).

REACTANCE, MECHANICAL X_m **(BS)**

Definition: The magnitude of the imaginary part of the mechanical impedance Z_m. $Z_m = R_m + i X_m$ (R_m = mechanical resistance).
Unit: Newton second per metre. $M T^{-1}$ (vector).
Note: If the reactance is solely due to inertia it is termed the mechanical mass reactance; if it is solely due to stiffness it is termed the mechanical stiffness reactance.

REACTANCE, SPECIFIC ACOUSTICAL X_s **(BS)**

Definition: The magnitude of the imaginary part of the specific acoustical impedance Z_s. $Z_s = R_s + i X_s$ (R_s = specific acoustical resistance).
Unit: Pascal second per metre. $M L^{-2} T^{-1}$ (vector).
Note: (1) The quantity was formerly termed the unit-area acoustical reactance.
(2) If the reactance is solely due to inertia it is termed the specific acoustical mass reactance; if it is solely due to stiffness it is termed the specific acoustical stiffness reactance.

REDUCTION FACTOR, SOUND

See transmission coefficient (note 4)

REFLECTANCE

See reflexion coefficient (note 1)

REFLECTIVITY

Definition: The total reflexion coefficient of a layer of substance of such a thickness that the reflexion coefficient does not change for an increase of thickness. Any conditions imposed will be the same as those for reflexion coefficient (qv).
Unit: There are no units or dimensions (scaral).

REFLEXION COEFFICIENT

Symbol: ρ; sometimes r.
Definition: Energy E_r reflected by a surface divided by that incident on it (E_o) under identical conditions. $\rho = E_r/E_o$.
Unit: There are no units or dimensions (scalar).
Note: (1) The quantity is also termed the reflexion factor or reflectance, the use of the last named being deprecated.
(2) The reflexion coefficient is generally expressed as a percentage.
(3) $\rho + \delta + \tau = 1$, where δ = scattering coefficient and τ = transmission coefficient.
(4) In sound, the reflexion coefficient depends on the frequency of the sound.
(5) In light there are two reflexion coefficients; the direct reflexion coefficient is concerned with specular reflexion, ie reflexion in accordance with the usual laws of optics, where the angle of reflexion is equal to the angle of incidence; the diffuse reflexion coefficient is concerned with light reflected from every part of a surface in many directions. Both depend on the angle of incidence, the mode of illumination and the wavelength distribution of the light.

REFLEXION DENSITY

Definition: The common logarithm of the ratio of the luminance L of a non-absorbing perfect diffuser to that of the sample L_o, both being illuminated at an angle of forty-five degrees to the normal, the direction of measurement being normal to the surface. Reflexion density = log (L/L_o).
Unit: There are no units or dimensions (scalar).

REFRACTIVE INDEX, ABSOLUTE

Symbol: n; formerly μ.
Definition: Velocity of light in a vacuum c_o divided by the velocity of light in the medium concerned c. $n = c_o/c$. Alternatively, if the medium has an interface with the vacuum, the angle i between an incident ray of light in the vacuum and the normal is related to the angle r between the refracted ray in the medium and the normal according to Snell's law: $n = \sin i/\sin r$.
Unit: There are no units or dimensions (scalar).
Note: The relative refractive index is defined in the same way, but in respect of two media rather than one medium and a vacuum.

RELUCTANCE

Symbol: R; sometimes R_m or $\mathscr{R}$ or S (BS).
Definition: Magnetic potential difference U divided by magnetic flux Φ. $R = U/\Phi$.

Unit: Per henry. $M^{-1} L^{-2} T^2 I^2$; $L^{-1} \mu^{-1}$; $L T^{-2} \varepsilon$ (scalar).
Note: $(U/R) \overset{*}{R} = 4\pi R$; $\overset{*}{R} = \overset{*}{U}/\Phi$.

RESILIENCE

Definition: Work done W in distorting a body to some predetermined limit (eg its elastic limit or breaking point) divided by the volume V of the body. Resilience $= W/V$.
Unit: Joule per metre cubed. $M L^{-1} T^{-2}$ (scalar).

RESISTANCE, ACOUSTICAL R_a (BS)

Definition: The real part of the acoustical impedance Z_a. $Z_a = R_a + i X_a$ (X_a = acoustical reactance).
Unit: Pascal second per metre cubed. $M L^{-4} T^{-1}$ (vector).

RESISTANCE, ELECTRICAL R

(1) RESISTANCE TO A DIRECT CURRENT
   Definition: Electric potential difference V divided by current I when there is no electromotive force in a conductor. $R = V/I$.
(2) RESISTANCE TO AN ALTERNATING CURRENT
   Definition: The real part of the electrical impedance Z. $Z = R + i X$
   (X = electrical reactance).
Unit: Ohm. $M L^2 T^{-3} I^{-2}$; $L T^{-1} \mu$; $L^{-1} T \varepsilon^{-1}$ (scalar).

RESISTANCE, MECHANICAL

Symbol: R_m (BS); occasionally r (BS).
Definition: The real part of the mechanical impedance Z_m. $Z_m = R_m + i X_m$
(X_m = mechanical reactance).
Unit: Newton second per metre. $M T^{-1}$ (vector).

RESISTANCE, SPECIFIC ACOUSTICAL R_s (BS)

Definition: The real part of the specific acoustical impedance Z_s.
$Z_s = R_s + i X_s$ (X_s = specific acoustical reactance).
Unit: Pascal second per metre. $M L^{-2} T^{-1}$ (vector).
Note: The quantity was formerly called the unit-area acoustical resistance.

RESISTANCE, THERMAL R (BS)

(1) *Definition:* The reciprocal of thermal conductance C. $R = 1/C$.
   Unit: Metre squared kelvin per watt. $L^2 T Q^{-1} \Theta$; $M^{-1} L^2 T$ (scalar).

(2) *Definition:* Temperature difference θ between two points divided by the mean rate of flow of entropy S with respect to time t. $R = \theta \left/ \dfrac{dS}{dt} \right.$

Unit: Thermal ohm = kelvin squared per watt. $T\ Q^{-1}\ \Theta^2$; $M^{-1}\ L^2\ T^{-1}$ (scalar).

Note: The former definition was: the temperature difference between two points divided by the rate of flow of heat energy Q with respect to time.

$R = \theta \left/ \dfrac{dQ}{dt} \right.$

RESISTIVITY ρ

Definition: Electric field strength E divided by current density J. $\rho = E/J$. Alternatively, the resistivity is the product of the electrical resistance R and the cross-sectional area A, divided by the length l. $\rho = RA/l$.
Unit: Ohm metre. $M\ L^3\ T^{-3}\ I^{-2}$; $L^2\ T^{-1}\ \mu$; $T\ \varepsilon^{-1}$ (scalar).
Note: The quantity is also termed the specific resistance.

RESISTIVITY, MASS ρ_m **(BS)**

Definition: The product of resistivity ρ and density d. $\rho_m = \rho d$.
Unit: Ohm kilogramme per metre squared. $M^2\ T^{-3}\ I^{-2}$; $M\ L^{-1}\ T^{-1}\ \mu$; $M\ L^{-3}\ T\ \varepsilon^{-1}$ (scalar).

RESISTIVITY, THERMAL ϕ

Definition: The reciprocal of the thermal conductivity λ. $\phi = 1/\lambda$.
Unit: Metre kelvin per watt. $L\ T\ Q^{-1}\ \Theta$; $M^{-1}\ L\ T$ (scalar).
Note: The quantity is also termed the thermal fluidity.

RESTITUTION COEFFICIENT e **(O)**

Definition: Velocity v_2 of a body at the instant immediately following an impact divided by the velocity v_1 at the instant immediately preceding the impact. $e = v_2/v_1$.
Unit: There are no units or dimensions (scalar).

RIGIDITY MODULUS

See elastic modulus (2)

SATURATION RATIO ψ

Definition: Actual specific humidity x divided by the specific humidity x_0 of saturated air at the same temperature. $\psi = x/x_0$.
Unit: There are no units or dimensions (scalar).

SCATTERING COEFFICIENT δ

Definition: Energy E_s scattered by a surface or medium divided by that incident at the point (E_o) under identical conditions. $\delta = E_s/E_o$.
Unit: There are no units or dimensions (scalar).
Note: (1) The quantity is also termed the scattering factor or dissipation coefficient (or factor).
(2) $\delta = a - \tau$, where a = absorption coefficient and τ = transmission coefficient.

SCOTOPTIC . . .

All luminous quantities as defined may apply to the scotoptic system of photometry, ie with the human eye fully dark-adapted. The adjective *scotoptic* precedes the name of the quantity.

SECTION MODULUS

Symbol: Z or W.
Definition: The section modulus of a plane area about an axis in its plane is the ratio of the second moment of area I to the distance r_o from the axis to the most remote point of the area. $Z = I/r_o$.
Unit: Metre cubed. L^3 (scalar).

SEEBECK COEFFICIENT a_S

Definition: Difference between the heat energy ΔE liberated at the cold junction of a thermocouple and that absorbed at the hot junction, divided by the charge Q flowing through the thermocouple. $a_s = \Delta E/Q$.
Unit: Joule per coulomb = volt. $M\ L^2\ T^{-3}\ I^{-1}$; $M^{\frac{1}{2}}\ L^{3/2}\ T^{-2}\ \mu^{\frac{1}{2}}$; $M^{\frac{1}{2}}\ L^{\frac{1}{2}}\ T^{-1}\ \varepsilon^{-\frac{1}{2}}$ (scalar).

SENSATION LEVEL

Symbol: L or L_p; sometimes S (O).
Definition: The common logarithm of the ratio of the intensity I or power P or energy E of a sound to the threshold level of aural sensation, I_o or P_o or E_o respectively, the two being expressed in the same units.
$L = \log (I/I_o) = \log (P/P_o) = \log (E/E_o)$.
Unit: Bel; in practice the decibel is always used. There are no dimensions (scalar).
Note: (1) The experimentally-determined values of the threshold level (for binaural hearing of the standard sinusoidally-varying plane progressive tone of frequency one thousand hertz) are $I_o = 2 \cdot 5 \times 10^{-12}$ watts per metre squared, $P_o = 2 \cdot 5 \times 10^{-12}$ watts, $E_o = 2 \cdot 5 \times 10^{-12}$ joules.

(2) The sensation level is also equivalent to twice the common logarithm of the ratio of the root-mean-square pressure p to the threshold level of aural sensation p_0, each expressed in the same units. $L = 2 \log (p/p_0)$. The experimentally-determined value of p_0 is $3 \cdot 16 \times 10^{-5}$ pascals.

(3) See the entry intensity level. The intensity level of the threshold sensation level is experimentally determined as $3 \cdot 98$ decibels.

SHEAR, ANGLE OF

See strain (2) (note)

SHEAR MODULUS

See elastic modulus (2)

SPECIFIC HEAT CAPACITY (MASS BASIS)

Symbol: c. Under the conditions of constant pressure or of constant volume the symbols are c_p and c_v respectively.

Definition: Heat capacity $dQ/d\theta$ divided by mass m. $c = \dfrac{1}{m} \dfrac{dQ}{d\theta}$ (Q = quantity of heat, θ = temperature).

Unit: Joule per kilogramme kelvin. $M^{-1} Q \Theta^{-1}$; no mechanical dimensions (scalar).

Note: The older name for the quantity, specific heat, is deprecated.

SPECIFIC HEAT CAPACITY, VOLUME BASIS

Definition: Heat capacity $dQ/d\theta$ divided by volume V. Specific heat capacity, volume basis $= \dfrac{1}{V} \dfrac{dQ}{d\theta}$ (Q = quantity of heat, θ = temperature).

Unit: Joule per metre cubed kelvin. $L^{-3} Q \Theta^{-1}$; $M L^{-3}$ (scalar).

SPEED

See velocity (note)

SPEED, ROTATIONAL

See frequency, rotational (note)

STERADIANCY

See radiance (note)

STIFFNESS, ACOUSTICAL S_a (BS)

Definition: The product of the acoustical stiffness reactance X_a and the angular frequency ω. $S_a = X_a\omega$. If the volume V of an enclosure has dimensions that are small in comparison with the wavelengths involved the acoustical stiffness is the product of the density ρ of the medium and the square of the velocity c of the wave propagation, divided by the volume. $S_a = \rho\, c^2/V$.
Unit: Pascal per metre cubed. $M\ L^{-4}\ T^{-2}$ (vector).

STIFFNESS (MECHANICAL) s

Definition: The product of the mechanical stiffness reactance X_m and the angular frequency ω. $s = X_m\omega$.
Unit: Newton per metre. $M\ T^{-2}$ (vector).

STRAIN

(1) TENSILE STRAIN
   Symbol: e; sometimes ε.
   Definition: The fractional increase in length Δl from the length l_0 in a specified reference state. $e = \Delta l/l_0$.
   Note: The quantity is also termed the linear strain or fractional (or relative) elongation.
(2) SHEAR STRAIN
   Symbol: γ; sometimes ϕ (BS).
   Definition: The shear strain is given by the angle (in radians) through which twisting takes place.
   Note: The quantity is also termed the angle of shear (shear angle).
(3) BULK STRAIN θ
   Definition: The fractional increase in volume ΔV from the volume V_0 in a specified reference state. $\theta = \Delta V/V_0$.
   Note: The quantity is also called the volume strain or hydrostatic strain.
Unit: There are no units or dimensions (except that shear strain may be measured in radians) (scalar).

STRENGTH (OF A SOURCE), SOUND A (BS)

Definition: Rate of volume V displacement of the surface which constitutes the source. $A = dV/dt$ (t = time).
Unit: Metre cubed per second. $L^3\ T^{-1}$ (scalar).

STRESS

See pressure (note)

SURFACE TENSION

Symbol: γ; sometimes σ.
Definition: Force F across a line element in a surface divided by the length l of the line element. $\gamma = F/l$. Alternatively, surface tension is the free (ie isothermal) surface energy E divided by the area A over which the energy is applied. $\gamma = E/A$.
Unit: Newton per metre (equivalent to the joule per metre squared). $M\ T^{-2}$ (vector).

SUSCEPTANCE B

Definition: The magnitude of the imaginary part of the admittance Y. $Y = G + i B$ (G = electrical conductance).
Unit: Siemens. $M^{-1}\ L^{-2}\ T^3\ I^2$; $L^{-1}\ T\ \mu^{-1}$; $L\ T^{-1}\ \varepsilon$ (scalar).

SUSCEPTIBILITY, ELECTRIC χ_e

Definition: One less than the relative permittivity ε_r. $\chi_e = \varepsilon_r - 1$.
Unit: There are no units or dimensions (scalar).
Note: (U/R) $\overset{*}{\chi}_e = \chi_e/4\pi$; $\overset{*}{\chi}_e = (\varepsilon_r - 1)/4\pi$.

SUSCEPTIBILITY, MAGNETIC κ

Definition: One less than the relative permeability μ_r. $\kappa = \mu_r - 1$.
Unit: There are no units or dimensions (scalar).
Note: (1) The quantity is more correctly termed the magnetic volume susceptibility. The magnetic mass susceptibility χ (BS) is the volume susceptibility divided by the density ρ. $\chi = \kappa/\rho$. This latter quantity is also termed the specific susceptibility.
(2) (U/R) $\overset{*}{\kappa} = \kappa/4\pi$; $\overset{*}{\kappa} = (\mu_r - 1)/4\pi$. $\overset{*}{\chi} = \chi/4\pi$; $\overset{*}{\chi} = \overset{*}{\kappa}/\rho$.

TEMPERATURE

(1) ABSOLUTE, ie THERMODYNAMIC TEMPERATURE
   Symbol: T; sometimes Θ.
   Definition: There is no formal definition of absolute temperature. It can only be explained in terms of an inadequate synonym, eg degree of hotness.
   Unit: Kelvin. Θ; $L^2\ T^{-2}$ (scalar).
(2) CUSTOMARY TEMPERATURE
   Symbol: θ; sometimes t.
   Definition: The difference between the absolute temperature T and a conventional constant T_0. $\theta = T - T_0$. (The value of T_0 is 273·15 kelvins.)
   Unit: Degree Celsius. Θ; $L^2\ T^{-2}$ (scalar).

(3) TEMPERATURE INTERVAL or DIFFERENCE

Symbol: θ; sometimes t. Also $\Delta\theta$, ΔT; sometimes Δt, $\Delta\ominus$.
Definition: The difference between two temperatures θ_1, θ_2, each measured in the same units. $\theta = \theta_1 - \theta_2$.
Unit: Kelvin. $\ominus$; $L^2 T^{-2}$ (scalar).
Note: The quantity is also termed the thermal potential difference, and the unit is then termed the thermal volt.

Note: (1) The several different kinds of temperature include the following examples.

(a) Colour temperature (of a light source). The temperature of a full (ie Planckian) radiator that would emit radiation of substantially the same spectral distribution in the visible region as the radiation from the source, and which would produce in the eye the same sensation of colour as the given radiator. An alternative definition implies the same colour sensation in the eye but not necessarily the same spectral distribution. This alternative definition is not preferred, and the temperature as defined by it is not identical with that in the first case.

(b) Luminance temperature (of a light source at a given wavelength). The temperature of a full (ie Planckian) radiator which has the same luminance as the radiator concerned at the given wavelength. This was formerly termed the brightness temperature.

(2) Because of the difficulty of measuring the absolute temperature of a body, a practical scale has been devised such that temperatures on it agree as closely as possible with those of the thermodynamic scale (and certainly within the limits of experimental determination). The International Practical Temperature Scale of 1968 (IPTS-68) is based on the concept of assigning values to certain reproducible equilibrium states (fixed points), and on standard interpolation procedures. The defining fixed points are given in the following table.

state	assigned value on the IPTS-68		estimated uncertainty kelvin
	kelvin	°C	
equilibrium between the solid, liquid and vapour phases of equilibrium hydrogen[1] (triple point of equilibrium hydrogen)[2]	13·81	−259·34	0·01
equilibrium between liquid equilibrium hydrogen[1] and its vapour at a pressure of 25/76 standard atmosphere (33 330·6 pascals)	17·042	−256·108	0·01
equilibrium between liquid equilibrium hydrogen[1] and its vapour (boiling point of equilibrium hydrogen)[3]	20·28	−252·87	0·01

state	assigned value on the IPTS–68		estimated uncertainty kelvin
	kelvin	°C	
equilibrium between liquid neon and its vapour (boiling point of neon)[3]	27·102	−246·048	0·01
equilibrium between the solid, liquid and vapour phases of oxygen (triple point of oxygen)[2]	54·361	−218·789	0·01
equilibrium between liquid oxygen and its vapour (boiling point of oxygen)[3]	90·188	−182·962	0·01
equilibrium between the solid, liquid and vapour phases of water (triple point of water)[2, 4]	273·16[5]	0·01[5]	0[6]
equilibrium between liquid water and its vapour (boiling point of water)[3, 4] or	373·15	100	0·005
equilibrium between solid and liquid tin (freezing point of tin)[3]	505·118 1	231·968 1	0·015
equilibrium between solid and liquid zinc (freezing point of zinc)[3]	692·73	419·58	0·03
equilibrium between solid and liquid silver (freezing point of silver)[3]	1 235·08	961·93	0·2
equilibrium between solid and liquid gold (freezing point of gold)[3]	1 337·58	1 064·43	0·2

Notes to the table

The interpolation procedures employed are:

−259·35 °C to 630·74 °C, measurement of resistance using a platinum resistance thermometer;

630·74 °C to 1 064·43 °C, measurement of electromotive force using a platinum/10% platinum-90% rhodium thermocouple;

above 1 064·43 °C, measurement of spectral concentration of radiation and the application of Planck's black-body law.

[1] Ie hydrogen having its equilibrium ortho-para composition
[2] At zero pressure
[3] At standard atmospheric pressure (101 325 pascals)
[4] With the isotopic composition of sea water
[5] Values fixed by definition
[6] By definition

TEMPERATURE COEFFICIENT α

Definition: Temperature coefficient is defined in connexion with a quantity X that varies with temperature θ. Often, for quantities that vary linearly (or approximately so) with temperature, the coefficient is defined as the fractional increase in X divided by the increase in temperature under specified conditions.

$$\alpha = \frac{\Delta X}{X\,\Delta\theta} \approx \frac{1}{X}\frac{dX}{d\theta}.$$

Unit: Per kelvin. Θ^{-1}; $L^{-2}\,T^2$ (scalar).

Note: Expansion coefficient (qv) and pressure coefficient (qv) are examples of temperature coefficient.

TEMPERATURE GRADIENT

Definition: The rate of change of temperature θ with respect to distance x. Temperature gradient $= d\theta/dx$.

Unit: Kelvin per metre. $L^{-1}\,\Theta$; $L\,T^{-2}$ (scalar).

TENSION

See force (note 1)

TENSION, ELECTRIC

See potential difference, electric (note)

TENSION, MECHANICAL

See pressure (note)

THOMSON COEFFICIENT

Symbol: σ; sometimes α_T.

Definition: Heat energy E developed between two points in a wire as a result of a difference in temperatures divided by the product of the charge Q flowing and the difference in temperature ΔT between the two points.

$\sigma = E/Q\,\Delta T.$

Unit: Joule per coulomb kelvin = volt per kelvin. $M\,L^2\,T^{-3}\,I^{-1}\,\Theta^{-1}$; $M^{\frac{1}{2}}\,L^{3/2}\,T^{-2}\,\mu^{\frac{1}{2}}\,\Theta^{-1}$; $M^{\frac{1}{2}}\,L^{\frac{1}{2}}\,T^{-1}\,\varepsilon^{-\frac{1}{2}}\,\Theta^{-1}$ (scalar).

Note: (1) This coefficient measures the Thomson effect, and the definition assumes that the Joule effect (the heat developed in a conductor as a result of the flow of current) is zero.

(2) The Thomson coefficient was called the specific heat of electricity by William Thomson (later Lord Kelvin).

THRUST

See force (note 1)

TIME t

Definition: There is no formal definition of time. It can only be explained in terms of a synonym, eg duration.
Unit: Second. T (scalar).
Note: (1) The symbol T is used for periodic time (period), ie time taken to describe one complete cycle of operations.
(2) The symbol τ (occasionally T) is used for the time constant of an exponentially varying quantity $f(t)$, and is defined as the time after which the quantity would reach a specified limit if it maintains its initial rate of variation. $f(t) = A + B\,e^{-t/\tau}$.
(3) The symbol T is used in acoustics for reverberation time: this is the time required for the average sound energy density in an enclosure, initially in a steady state, to decrease to one-millionth of its initial value after the sound has ceased. This decrease represents a fall of 60 decibels in intensity level.

TRAFFIC FACTOR

(1) FUEL CONSUMPTION F (O)
   Definition: Volume V of fuel used up in travelling a certain distance s, divided by the distance. $F = V/s$.
   Unit: Metre squared. L^2 (scalar).
(2) MASS-DISTANCE
   Definition: Mass m carried multiplied by distance s travelled. Mass-distance $= ms$.
   Unit: Kilogramme metre. M L (vector).
(3) MASS PER FUEL CONSUMPTION
   Definition: Mass m carried divided by fuel consumption F.   Mass per fuel consumption $= m/F$.
   Unit: Kilogramme per metre squared. M L^{-2} (scalar).

TRANSFER COEFFICIENT, HEAT

See conductance, thermal (note)

TRANSMISSION COEFFICIENT τ

Definition: Energy E_t transmitted through and beyond a surface divided by that incident on it (E_0) under identical conditions. $\tau = E_t/E_0$.
Unit: There are no units or dimensions (scalar).
Note: (1) The quantity is also termed the transmission factor or transmittance, the use of the last named being deprecated.

(2) The transmission coefficient is generally expressed as a percentage.

(3) $\tau + \rho + \delta = 1$, where ρ = reflexion coefficient and δ = scattering coefficient.

(4) In sound, the transmission coefficient depends on the frequency of the sound. Also, $R = 10 \log (1/\tau)$, where R is the sound reduction factor or sound reduction index.

(5) In light there are two transmission coefficients: the direct transmission coefficient is concerned with transmission without scatter; the diffuse transmission coefficient is concerned with light transmitted with scatter in many directions. Both depend on the angle of incidence, the mode of illumination and the wavelength distribution of the light. Also, the optical density $= \log (1/\tau)$.

TRANSMISSIVITY

Definition: The internal transmission coefficient of unit thickness of a transmitting material under conditions in which the boundary of the material has no influence.

Unit: There are no units or dimensions (scalar).

TRANSMITTANCE

See transmission coefficient (note 1); transmittancy (note)

TRANSMITTANCE, THERMAL

See conductance thermal (note)

TRANSMITTANCY

Definition: Transmission coefficient of a liquid or solid solution divided by that of the solvent of the same form and thickness.

Unit: There are no units or dimensions (scalar).

Note: The quantity has also been known by the deprecated name transmittance.

VAPOUR CONCENTRATION

See humidity, absolute (note)

VAPOUR DENSITY

See density, relative (note 1)

VECTOR POTENTIAL, MAGNETIC A

Definition: The curl of the magnetic vector potential is equal to the magnetic flux density B. $\nabla \times A = B$.

Unit: Weber per metre. $M\,L\,T^{-2}\,I^{-1}$; $M^{\frac{1}{2}}\,L^{\frac{1}{2}}\,T^{-1}\,\mu^{\frac{1}{2}}$; $M^{\frac{1}{2}}\,L^{-\frac{1}{2}}\,\varepsilon^{-\frac{1}{2}}$ (vector).

VELOCITY

Symbol: v. The symbols u, v, w are used for the vector components of velocity. The symbol c is used for the velocity of electromagnetic radiation in free space; c_0 is used if c must be kept for phase velocity in a medium. In problems on translational motion u is used for initial velocity.
Definition: Rate of change of displacement s with respect to time t. $v = ds/dt$.
Unit: Metre per second. L T⁻¹ (vector).
Note: The term speed refers to the scalar form of velocity, ie linear or translational velocity.

VELOCITY, ANGULAR

Symbol: ω; sometimes Ω.
Definition: Rate of change of angle θ with respect to time t. $\omega = d\theta/dt$.
An alternative definition is: one circle 2π (when expressed in radians) multiplied by frequency f. $\omega = 2\pi f$.
Unit: Radian per second, degree per second. T⁻¹ (vector).
Note: (1) The term angular velocity implies reference to analytical angle; an alternative term, rotational velocity, implies reference to geometrical angle, although there is no practical distinction between the two. Other alternative terms are angular frequency, circular frequency and pulsitance, the use of the last named being deprecated.
(2) This quantity must be distinguished from rotational frequency.

VELOCITY GRADIENT

See viscosity

VELOCITY POTENTIAL ϕ (BS)

Definition: The gradient of the velocity potential is equal to the particle velocity v with a reversed sign. $\nabla\phi = -v$.
Unit: Per second. T⁻¹ (scalar).

VELOCITY RATIO V (O)

Definition: The velocity v_2 given to the effort (input force) of a machine divided by the velocity v_1 that the load (output force) acquires. $V = v_2/v_1$.
Unit: There are no units or dimensions (scalar).

VISCOSITY

Symbol: η; sometimes μ.
Definition: Tangential stress τ in a liquid undergoing streamline (ie laminar) flow divided by the velocity gradient. (The velocity gradient is the rate of

change of velocity v with distance z across the flow of liquid, dv/dz.)

$$\eta = \tau \bigg/ \frac{dv}{dz}.$$

Unit: Newton second per metre squared. $M L^{-1} T^{-1}$ (scalar).
Note: The quantity is more correctly termed the dynamic or absolute viscosity.

VISCOSITY, KINEMATIC v

Definition: Dynamic viscosity η divided by density ρ. $v = \eta/\rho$.
Unit: Metre squared per second. $L^2 T^{-1}$ (scalar).

VISCOSITY, SPECIFIC

Definition: Viscosity η_1 of a substance divided by the viscosity η_2 of a reference substance under conditions specified for both substances (normally equal temperatures). Specific viscosity $= \eta_1/\eta_2$.
Unit: There are no units or dimensions (scalar).
Note: In the case of liquids the reference substance is usually water.

VOLTAGE

See potential difference, electric (note)

VOLUME

Symbol: V; occasionally v.
Definition: The space contained by a closed area. The size of the volume will be expressed by a formula the nature of which depends on the shape of the space.
Unit: Metre cubed. L^3 (scalar).
Note: The term capacity is also used, often where the unit of measurement is not of the form cubic . . ..

VOLUME FLOW RATE

Symbol: q_v; sometimes U or q.
Definition: Rate of change of volume V with respect to time t. $q_v = dV/dt$.
Unit: Metre cubed per second. $L^3 T^{-1}$ (scalar).
Note: The quantity is also termed the volume velocity.

VOLUME, MOLAR

Definition: Volume V divided by molar value n. Molar volume $= V/n$.
Unit: Metre cubed per mole. $L^3 N^{-1}$ (scalar).
Note: N is used as the dimensional symbol of molar value.

176 Volume, Specific

VOLUME, SPECIFIC v

Definition: Volume V divided by mass m, ie the reciprocal of density. $v = V/m$.
Unit: Metre cubed per kilogramme. $M^{-1} L^3$ (scalar).

WAVELENGTH

See length (note)

WAVE NUMBER

Symbol: σ. The symbol $\bar{v}$ is used in spectroscopy.
Definition: The reciprocal of length l, especially wavelength λ. $\sigma = 1/l$, especially $\bar{v} = 1/\lambda$.
Unit: Per metre L^{-1} (generally regarded as scalar).
Note: The term circular wave number is used for the quantity k defined by $k = 2\pi\sigma$.

WEIGHT

See force (note 2)

WEIGHT DENSITY

See weight, specific (note 1)

WEIGHT, SPECIFIC γ

Definition: Weight W divided by volume V. $\gamma = W/V$.
Unit: Kilogramme-weight per metre cubed; kilogramme-force per metre cubed. $M L^{-2} T^{-2}$ (vector).
Note: (1) The quantity is also termed the weight density.
(2) Because the specific weight of a body has a value which depends on the value of the local acceleration of free fall, it is better to measure it in practice with reference to the standard value of the acceleration of free fall, or (best of all) in terms of absolute force, ie in newtons per metre cubed.

WORK

See energy (note)

YOUNG'S MODULUS

See elastic modulus (1)

PART 3
APPENDICES

Appendix 1

SYSTEMS OF UNITS

1 Introduction

Every fundamental formula of physical science can be placed in one of three categories.

(a) Empirical laws

These are laws connecting certain quantities that appear, as a result of observations of the behaviour of the universe, to be true. They can be tested and shown to hold, often by laboratory experiment, but they cannot be validated absolutely from theoretical considerations. They are axioms from which it is possible to logically deduce the structure of the branch of science in which they are relevant. One example is Newton's second law of motion: the force F on a body is measured by the rate of change of momentum p it produces, and the change takes place in the direction of the force. This law is generally reduced to the equation

$$F \propto \frac{dp}{dt}$$

(t = time).

(b) Defining equations

These are equations set up for our convenience to connect certain quantities in such a manner as to provide a new quantity which is useful. Thus, in order to state Newton's second law of motion in the form given above we define momentum by the product of mass m and velocity v:

$$p \propto mv.$$

(c) Derived relationships

These are relationships between quantities deduced from the imposition of a set of conditions on a physical system. For example, the velocity achieved after a time t by a body that has moved with a uniform acceleration a from a state in which its velocity was u is given by

$$v \propto u + at.$$

The size of each of the quantities that occur in all three categories of formula is expressed as a numerical value together with a unit. If the units are chosen in an arbitrary fashion the constant of proportionality in each equation will take a value that depends on the units themselves, but if they are chosen in a systematic fashion, ie they belong to one of the recognised (or unrecognised but possible) systems of units, the constant takes a given fixed value that can be chosen at will.

In the following notes examples of the terms explained are given in square brackets.

2 Types of systems

Every system can be described as either coherent or non-coherent, but some systems can be grouped under other headings.

(a) Coherent (or unitary, or 1:1) systems

A system is described as coherent if a quantity of unit size is derived from the combination of fundamental or previously derived units, each being of unit size.

[The SI is coherent. Eg one unit of force (newton) is the force experienced by one unit of mass (kilogramme) acted on by one unit of acceleration (metre per second squared).]

(b) Non-coherent systems

A system which does not conform to the condition stated in 2(a) is non-coherent.

[The MkgfS system is non-coherent. Eg one unit of force (kilogramme-force) is the force experienced by one unit of mass (kilogramme) acted on by 9·806 65 units of acceleration (metres per second squared).]

(c) Gravitational systems

A system is described as gravitational if the size of one of its fundamental units (that of mass or force) depends on the size of g_n, the standard acceleration of free fall. It can be either coherent or non-coherent.

[The FSS system is a coherent gravitational system. The units of mass (slug) has a size of (g) pounds, where (g) is the numerical value of g_n, the latter being expressed in feet per second squared. The MkgfS system is a non-coherent gravitational system, as indicated in 2(b): $g_n = 9·806\ 65$ metres per second squared.]

(d) Technical systems

A system is described as technical if the definition of the unit of force involves an acceleration equal to that of the standard or local acceleration of free fall. Such a system is thus gravitational and non-coherent. The name must not be regarded as implying that the system is particularly recommended for technical purposes.

[The MkgfS system is a technical system: see 2(b), 2(c).]

(e) Practical systems

A system is described as practical if its units are all used in practice for standard types of measurements. A practical system is incomplete in that it employs only a selection of units; however, every system has its practical collection of units.

[The term practical has usually been applied to the MKSΩ system in which the units of basic electrical quantities were defined in terms of experimentally-measured quantities. The Hartree system is also a practical system.]

(f) Natural systems

A system is described as natural if its units are of a convenient size for the measurement of quantities associated with the atom.
[The Hartree system is a natural system.]

(g) Rationalised and unrationalised systems

A system is described as rationalised if the empirical laws and derived relationships that define its units are put in a logical, not irrational form. Thus relationships that are concerned with circular symmetry have a factor (coefficient) of 2π, and those concerned with spherical symmetry have a factor of 4π; other relationships do not contain the factor π. A system that does not conform to this condition is termed unrationalised. All systems may be put in a rationalised or unrationalised form, but the problem of rationalisation is normally confined to electrical relationships.
[The SI is made rationalised; the CGS system is normally unrationalised. In the former the capacitance C of a parallel-plate condenser is given by

$$C = \varepsilon A/d,$$

in the latter by

$$C = \varepsilon A/4\pi d$$

(ε = absolute permittivity of the medium between the plates, A = common area of the plates, d = separation of the plates, all the units in each case being those of the corresponding system).]

3 Types of units

(a) Fundamental (or basic) units

The sizes of certain units in each system are given formal definitions based as far as possible on naturally occurring conditions, and in such a manner as to be experimentally determined to an accuracy limited only by the apparatus and methods employed for their evaluation. These units are, where possible, totally unrelated to one another so that if it were found necessary to change the size or definition of one of the units (as has been done several times in the past) the sizes of the remaining units need not be altered. Such units are referred to as fundamental. One important aim is to produce a definitive system in which there are sufficient fundamental units to cover all possible relationships between quantities measured by the system.
The fundamental units normally employed in the various systems have been selected from those of the following quantities:

length
mass or force   (as yet the size of the unit of mass has not been defined in terms of a naturally-occurring condition, but of the mass of one particular object)

time
energy   (units of thermal and electrical energy have in the
                                past been defined independently of those of mecha-
                                nical energy)
electric current
    or resistance   (the former is now preferred)
temperature
luminous intensity
    molar value   (the formal definition awaits international considera-
                                tion)

The units of quantities other than length, mass, time and temperature are not fundamental in the sense that their sizes depend on the sizes of these four, but they are regarded as fundamental for reasons of convenience.

If, as in the case of mass, a certain standard has to be employed, the size of the standard need not be identical with the size of the fundamental unit of a given system.

[The mass standard is the International prototype kilogramme and the fundamental unit of mass in the CGS system is the gramme. 1 gramme = 0·001 kilogramme.]

(b) Derived units

Any unit, the size of which is determined directly or indirectly from the fundamental units of a system by means of an empirical law or defining equation, is described as a derived unit. By indirect determination it is implied that previously derived units or supplementary units are used instead of (or as well as) the fundamental units.

[The unit of momentum is a derived unit in most systems, being determined from the equation given in 1(b) together with an appropriate value for the constant of proportionality.]

The names of derived units can be classified as follows:

(i) Certain derived units have special names.

[In the SI the unit of force is called the newton.]

(ii) Many derived units have names which are combinations of the fundamental units, the names being indicative of their definitions.

[In the SI the unit of density, defined as mass divided by volume, is called the kilogramme per metre cubed.]

(iii) Some derived units have names which are combinations of other derived units (if necessary, together with fundamental units).

[In the SI the unit of torque is called the newton metre; it could also be called the kilogramme metre squared per second squared.]

(c) Supplementary units

The sizes of some units can be specified in a totally theoretical manner because the quantities they measure are free from any physical nature (ie are dimensionless), being defined in terms of the ratio of like physical quantities. These are described as supplementary units, and are required in addition to the fundamental units to specify certain derived units. Being dimensionless, they are of identical size in every system. The two supplementary units required are those of plane angle and of solid angle.

(d) Absolute units

A unit that belongs to a given system is described as abolute in that system if it is defined in a coherent, theoretical manner without reference to experimental determination.

[In the FSS system the unit of length (foot) is an absolute unit. The unit of mass (slug) is not, since a knowledge of its size can only be obtained through an experimental determination of the acceleration of free fall. The kilogramme is an absolute unit of mass in the SI but not in the CGS system, since in the latter system it is not coherent.]

(e) Arbitary units

An arbitrary unit does not belong to any of the recognised systems of units. [The ångström and the mile are arbitrary units of length. The calorie is an arbitary unit of heat energy except in the somewhat artificial CGS-thermal system.]

(f) International units

A unit is described as international if it is defined with reference to experimental determination, but belongs to a coherent system.

[The MKSΩ system originally possessed international units for electrical quantities and, through the unit of power, for non-electrical quantities.]

(g) Metric and imperial units

A unit is described as metric if it belongs to or is derived from the units of the SI or CGS (or similar) system, the constant of the defining equation being a multiple or submultiple of 10. A unit is described as imperial if it belongs to or is derived from the units of the FPS (or a similar) system, the constant of the defining equation being an arbitrary but conventionally accepted number. A unit common to both systems (eg the second) or a supplementary unit (eg the radian) is not generally considered either metric or imperial. Other units (those of electricity, etc) are regarded as metric if they are related in a system by multiples or submultiples of 10. A unit such as the grade (0·01 right angle) may be considered as metric, although it is strictly not part of the metric system.

4 Systems of units

Table 1 summarises the main features of the various systems of units that have been employed at one time or another. The diagram is an attempt at a simple classification.

Table 1: Unit systems

name of system	type of system[1]	selected units* length	mass	time	electric current	force	notes
METRIC SYSTEMS							
1 SI[2]	C R	m*	kg*	s*	A*	N	Units of absolute temperature: K* / luminous intensity: cd* / molar value: mol* (probably to be added). / Now the universally accepted system.
2 MKSA	C R	m*	kg*	s*	A*	N	Also called the Giorgi system (1950): became the SI in 1954, although the latter name was not formally adopted until 1959 (11th CGPM[3]). It was formerly unrationalised.
3 MKSΩ	C U	m*	kg*	s*	A	N	Unit of electrical resistance: Ω*. The system originally employed international units, with an experimentally defined ohm and volt. A practical system,[4] first proposed by Giorgi in 1900 until replaced by the MKSA system.
4 MKS	C —	m*	kg*	s*	(none)	N[5]	Precursor of the MKSΩ system: the original Giorgi system.
5 MkgfS	N GT R	m*	kg	s*	A	kgf*	Formerly used by engineers as an alternative to the SI.
6 CGS	C —	cm*	g*	s*	(none)	dyn	Introduced in 1873 by the British Association for the Advancement of Science on the advice of Kelvin, and generally accepted until replaced by the SI.
7 CGS-thermal	N —	cm*	g*	s*	(none)	dyn	Units of mechanical energy: erg / thermal energy: cal*. / Formerly used in heat analyses as an alternative to the CGS system.
8 CGSm[6]	C U	cm*	g*	s*	Bi*	dyn	See section 9–8.
9 CGS-emu[7]	C U	cm*	g*	s*	abA*[8]	dyn	See section 9–7.
10 CGSe[9]	C U	cm*	g*	s*	Fr/s*	dyn	See section 9–10.
11 CGS-esu[10]	C U	cm*	g*	s*	statA*[11]	dyn	See section 9–9.
12 Gaussian	C U	cm*	g*	s*	statA*[11] Fr/s*	dyn	Two alternative systems are given. See section 9–11. Also called the mixed, or symmetric system of units.

No. System	Symbols	Length	Mass	Time	Electric	Force	Notes
13 Heaviside-Lorentz	C R	cm*	g*	s*	statA*[11] / Fr/s	dyn	Two alternative systems are given. See section 9-12. The earliest rationalised system, introduced in 1883.
14 CgfS	N GT —	cm*	g	s*	(none)	gf*	Formerly used by engineers as an alternative to the CGS system.
15 MTS	C U	m*	t*	s*	A	sn	Formerly used in France (introduced in 1914).
16 MTS-thermal	N U	m*	t*	s*	A	sn	Units of mechanical energy: kJ thermal energy: th*. Formerly used by French engineers in heat analyses as an alternative to the MTS system.
17 Mie	C U	cm*	10^7 g*	s*	A	10^7 dyn	Unit of energy: J. A compromise between the CGS and MKS systems, formerly used occasionally in Germany.
18 mm-mg-s	C —	mm*	mg*	s*	(none)	μdyn	An early system proposed by Gauss and Weber, but never employed.
19 MGS	C —	m*	g*	s*	(none)	10^2 dyn	Introduced in 1863 by the BAAS on the advice of Kelvin and rarely used, being shortly replaced by the CGS system.
20 dm-kg-ds	C —	dm*	kg*	ds*	(none)	dyn	Suggested by Moon in 1891 but never employed.

ARBITRARY/METRIC-DERIVED SYSTEMS

No. System	Symbols	Length	Mass	Time	Electric	Force	Notes
21 Hartree[12]	C (R)						Introduced in 1927 for atomic analyses.[12] A practical natural system of units.
22 Quantum-electrodynamics[12]	C (R)						An alternative practical natural system of units.[12]
23 Ludovici[13]	C R						Proposed in 1956 but not employed.[13]

IMPERIAL SYSTEMS

No. System	Symbols	Length	Mass	Time	Electric	Force	Notes
24 FPS	C —	ft*	lb*	s*	(none)[14]	pdl	Formerly used in the UK and the US.
25 FPS-thermal	N —	ft*	lb*	s*	(none)	pdl	Units of mechanical energy: ft pdl thermal energy: Btu*. Formerly used in heat analyses as an alternative to the FPS system.
26 FSS	C G —	ft*	slug*	s*	(none)[14]	lbf*	Formerly used by engineers as an alternative to the FPS system.
27 FlbfS	N GT —	ft*	lb	s*	(none)	lbf*	Formerly used by engineers as an alternative to the FPS and FSS systems.
28 Stroud	N GT —	ft*	lb*	s*	(none)	Lb*	Introduced by Stroud in 1880. 1 Lb = 1 lbf.
29 ft-gr-s	C —	ft*	gr*	s*	(none)	gr ft/s²	A system that has rarely been used.

Notes to table 1.

* Fundamental units of the system are marked with an asterisk. The unit symbols employed are:

A	ampère	dm	decimetre	gr	grain	Lb	Pound (ie pound-force)	pdl	poundal
abA	abampère	ds	decisecond	J	joule	lbf	pound-force	s	second
Bi	biot	dyn	dyne	K	kelvin	m	metre	sn	sthène
Btu	British thermal unit	Fr	franklin	kg	kilogramme	mg	milligramme	statA	statampère
cal	calorie	ft	foot	kgf	kilogramme-force	mm	millimetre	t	tonne
cd	candela	g	gramme	kJ	kilojoule	mol	mole	th	thermie
cm	centimetre	gf	gramme-force	lb	pound	N	newton	Ω	ohm

siemens	(conductance—modern name of unit)
tesla	(magnetic flux density—modern name of unit)
volt	(voltage)
weber	(magnetic flux—modern name of unit)
watt	(power)

1   C = coherent; G = gravitational; N = non-coherent; R = rationalised; T = technical; U = unrationalised.
2   Système International d'Unités (International system of units).
3   Conférence Générale des Poids et Mesures (General conference of Weights and Measures).
4   The only electrical units employed:

ampère	(current)
coulomb	(charge)
farad	(capacitance)
henry	(inductance)
ohm	(resistance)
joule	(energy)

together with: (the siemens, tesla, volt, weber and watt listed above).

5   Not called by this name when the system was used as described.
6   CGS-magnetic.
7   CGS electromagnetic system of units.
8   = dyne$^{\frac{1}{2}}$ (mechanical equivalent).
9   CGS-electrostatic.
10   CGS electrostatic system of units.
11   = dyne$^{\frac{1}{2}}$ centimetre per second (mechanical equivalent).
12   For the fundamental and derived units, see section 10.
13   The fundamental units are:
the constant of universal gravitation, the charge on the electron, the magnetic constant, the electric constant.
See section 11.
14   See section 9–14.

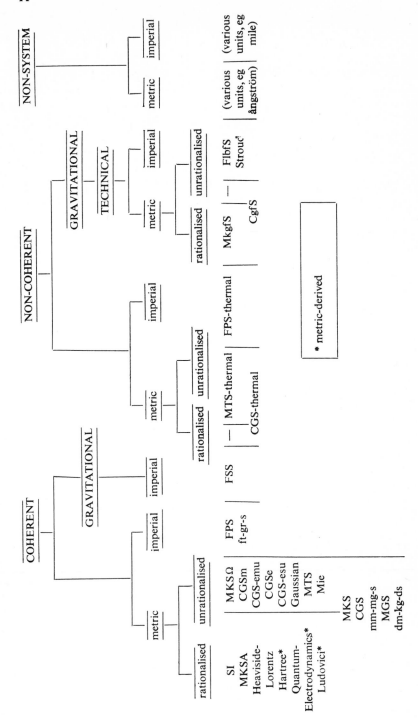

5 Mechanical quantities and units

5–1 Introduction

The following tables list all the fundamental and derived units, in each system, that have special names. Other derived units are named according to the rules given in section 3(*b*)(ii) and (iii). Derived units which, in some systems, have names that are combination forms are given in parentheses. Alternative names for some of the units will be found under the entries in part 1 (Units) for the names given here.

5–2 Systems: SI, MKSA, MKSΩ, MKS; CGS, CGS-thermal, C̈GSm, CGS-emu, CGSe, CGS-esu, Gaussian, Heaviside-Lorentz

A note to table 2 indicates that the name of the unit is not recognised internationally.

Table 2: Mechanical units (part 1)

	name of unit in given system	
quantity	SI, MKSA, MKSΩ, MKS	CGS, CGS-thermal, CGSm, CGS-emu, CGSe, CGS-esu, Gaussian, Heaviside-Lorentz
acceleration	(metre per second squared)	galileo[1]
action	planck[2]	(erg second)
energy	joule	erg
force	newton	dyne
frequency	hertz	hertz
length	metre	centimetre
mass	kilogramme	gramme
momentum	(kilogramme metre per second)	bole[3]
power	watt	(erg per second)
pressure	pascal[4]	barye[5]
reciprocal length	dioptre[6]	kayser[7]
time	second	second
velocity	(metre per second)	kine[8]
viscosity (dynamic)	(newton second per metre squared)	poise
viscosity, kinematic	(metre squared per second)	stokes
volume	stère[9]	(centimetre cubed)

Internationally accepted names:

1 centimetre per second squared　4 newton per metre squared　　7 per centimetre
2 joule second　　　　　　　　　5 dyne per centimetre squared　8 centimetre per second
3 gramme centimetre per second　6 per metre　　　　　　　　　9 metre cubed

5–3 MkgfS, MTS, MTS-thermal, MGS systems

Since these systems are based on the SI, the names of the units are identical

with or based on the names of those in the SI. In table 3 an asterisk marks
those units that are identical.

Table 3: Mechanical units (part 2)

quantity	name of unit in given system		
	MkgfS	MTS, MTS-thermal	MGS
energy	kilogrammetre	(metre sthène)	(gramme metre squared per second squared)
force	kilogramme-force	sthène	(gramme metre per second squared)
frequency	*hertz	*hertz	*hertz
length	*metre	*metre	*metre
mass	*kilogramme	tonne	gramme
pressure	(kilogramme-force per metre squared)	pièze	(gramme per metre second squared)
reciprocal length	*dioptre	*dioptre	*dioptre
time	*second	*second	*second
volume	*stère	*stère	*stère

5–4 dm-kg-ds system

This system shares only two units with the SI, those of mass (kilogramme)
and energy (joule); it shares no units with the CGS system.

5–5 CgfS, mm-mg-s systems

Since these systems are based on the CGS system, the names of the units are
identical with or based on the names of those in the CGS system. In table 4
a dagger marks those units that are identical.

Table 4: Mechanical units (part 3)

quantity	name of unit in given system	
	CgfS	mm-mg-s
acceleration	†galileo	(millimetre per second squared)
force	gramme-force	(milligramme millimetre per second squared)
frequency	†hertz	†hertz
length	†centimetre	millimetre
mass	†gramme	milligramme
pressure	(gramme-force per centimetre squared)	†barye
reciprocal length	†kayser	(per millimetre)
time	†second	†second
velocity	†kine	(millimetre per second)
viscosity (dynamic)	(gramme-force second per centimetre squared)	†poise

5–6 Mie system

The Mie system is a hybrid of the SI and CGS system, the names of the units being a combination of the units of the two systems. In table 5 an asterisk marks those units identical with SI units and a dagger those identical with CGS-system units.

Table 5: Mechanical units (part 4)

quantity	name of unit in Mie system
acceleration	†galileo
action	*planck
energy	*joule
frequency	*†hertz
length	†centimetre
mass	(10^4 kilogrammes)
power	*watt
reciprocal length	†kayser
time	*†second
velocity	†kine
viscosity, kinematic	†stokes

5–7 FPS, FPS-thermal, FSS, FlbfS, Stroud, ft-gr-s systems

Only the units of force, length, mass and time have received general recognition. NB: The Pound (spelled with a capital letter) is the unit of force equal in size to a pound-force; the pound (with a small initial letter) is a unit of mass.

Table 6: Mechanical units (part 5)

quantity	name of unit in given system				
	FPS, FPS-thermal	FSS	FlbfS	Stroud	ft-gr-s
acceleration	celo	celo	celo	celo	celo
energy	(foot poundal)	(foot pound-force)	duty	(foot Pound)	(grain foot squared per second squared)
force	poundal	(pound-force)	(pound-force)	Pound	(grain foot per second squared)
length	foot	foot	foot	foot	foot
mass	pound	slug	pound	pound	grain
time	second	second	second	second	second
viscosity (dynamic)	reyn	(pound-force second per foot squared)	(pound-force second per foot squared)	(Pound second per foot squared)	(grain per foot second)
volume flow rate	cusec	cusec	cusec	cusec	cusec

6 Heat quantities and units

In or associated with each system of units the units of measurement of heat quantities derive from the units of temperature and heat energy. In most systems the units of heat and other forms of energy are identical. These units are given in table 7.

Table 7: Basic thermal units

	name of unit of given quantity		
system	heat energy	absolute temperature	temperature difference
SI, MKSA, MKSΩ, MKS, Mie, dm-kg-ds	joule		
MkgfS	metre kilogramme-force		
CGS, CGSm, CGS-emu, CGSe, CGS-esu, Gaussian, Heaviside-Lorentz	erg		
CGS-thermal	calorie	kelvin	kelvin
CgfS	centimetre gramme-force		
MTS	metre sthène		
MTS-thermal	thermie		
mm-mg-s	milligramme millimetre squared per second squared		
MGS	gramme metre squared per second squared		
FPS	foot poundal		
FPS-thermal	British thermal unit	degree Fahrenheit (°F)	Fahrenheit degree (degF)
FSS, FlbfS	foot pound-force		
Stroud	foot Pound		
ft-gr-s	grain foot squared per second squared		

7 Photometric quantities and units

Table 8 lists the only units of photometric quantities that have special names in or associated with each system of units.

Table 8: Photometric units

	name of unit in given system				
quantity	SI, MKSA, MKSΩ, MKS, MkgfS, MTS, MTS-thermal, MGS	CGS, CGS-thermal, CGSm, CGS-emu, CGSe, CGS-esu, Gaussian, Heaviside-Lorentz, CgfS, Mie	mm-mg-s	dm-kg-ds	FPS, FPS-thermal, FSS, FlbfS, Stroud, ft-gr-s

Table 8—continued

	name of unit in given system				
illumination	lux	phot	lumen per millimetre squared	lumen per decimetre squared	lumen per foot squared
luminance[1]	nit	stilb	candela per millimetre squared	candela per decimetre squared	candela per foot squared
	apostilb	lambert	—	—	foot-lambert
luminous energy[2]	talbot	lumberg	(talbot)	(talbot)	(talbot)
luminous flux	lumen	lumen	lumen	lumen	lumen
luminous intensity	candela	candela	candela	candela	candela
luminous power	light-watt	light-watt	light-watt	light-watt	light-watt

1 The second-named unit is $1/\pi$ times the size of the first-named unit
2 The talbot and the lumberg are identical in size

8 Acoustical quantities and units

The only specially-named units in acoustics fall into one of the following groups.

(a) In the CGS, CGS-thermal, CGSm, CGS-emu, CGSe, CGS-esu, Gaussian and Heaviside-Lorentz systems:
the acoustical ohm measures acoustical resistance, reactance and impedance; the rayl (specific acoustical ohm) measures specific acoustical resistance, reactance and impedance.

(b) In the FPS, FPS-thermal, FSS, FlbfS, Stroud and ft-gr-s systems:.
the sabin measures equivalent absorption (area) of a surface.

(c) The units in table 9 are dimensionless, and therefore identical in all systems. Those of general level are also employed in non-acoustical circumstances.

Table 9: Dimensionless acoustical units

quantity	unit
equivalent loudness level (general)	phon
	bel (decibel), brig, neper
loudness	sone
noisiness	noy
perceived noise level	perceived noise decibel
pitch interval	cent, octave, savart, modified savart
subjective pitch	mel

9 Electrical and magnetic quantities and units

9–1 Introduction

Because of the complications that arise in the differences between the SI and the CGS system, and in the employment of rationalised or unrationalised formulae, the units of electricity and magnetism have been dealt with in great detail.

9–2 Fundamental formulae

(a) The force between current-carrying conductors

When currents I_1, I_2 are flowing along two parallel conductors at a separation x, a mechanical force per unit length F/l acting between them is observed:

$$\frac{F}{l} \propto \frac{I_1 I_2}{x}.$$

The theoretical deduction of this law from the Ampère-Laplace theorem for an infinitely-long straight conductor of negligible cross section shows that

$$\frac{F}{l} \propto \frac{2\mu I_1 I_2}{x} \qquad \cdots \qquad \cdots \qquad \cdots \qquad \cdots \qquad (1)$$

The constant 2 results from the integration involved in applying the theorem. μ is the absolute permeability of the medium in which the wires are situated, and is numerically equal to the ratio of the force which acts when the medium is present to that which acts in a vacuum, other conditions being unchanged, multiplied by a mathematical constant the size of which depends on the system of units.

(b) Coulomb's law of electrostatic forces

When charges Q_1, Q_2 are placed at a separation x a mechanical force F acting between them is observed:

$$F \propto \frac{Q_1 Q_2}{x^2}.$$

If the charges are situated in a medium of absolute permittivity ε Coulomb showed by experiment that:

$$F \propto \frac{Q_1 Q_2}{\varepsilon x^2} \qquad \cdots \qquad \cdots \qquad \cdots \qquad \cdots \qquad (2)$$

ε is numerically equal to the ratio of the force which acts in a vacuum to that which acts when the medium concerned fills the vacuum, other conditions being unchanged, multiplied by a constant the size of which depends on the system of units.

(c) The connexion between I and Q is given by the fundamental equation

$$I = \frac{dQ}{dt}$$

(t = time).

9–3 Rationalised and unrationalised formulae

Without a system of units other than that which applies to mechanical quantities the sizes of the constants of proportionality in equations (1) and (2) may be chosen at will, and appropriate units allotted to the quantities I, μ, Q, ε. If the constants of proportionality are each made equal to unity, the formulae and those derived from them are termed unrational. If the constants of proportionality are each made equal to $1/4\pi$ the formulae and those derived from them are termed rational. In the rationalised formulae the presence of a 4π implies a case of spherical symmetry and the presence of a 2π implies circular symmetry; this does not follow in unrationalised formulae. It should be noted that many formulae have identical rationalised and unrationalised forms.

9–4 The connexion between permeability and permittivity

Maxwell's electromagnetic theory showed that the absolute permeability of free space (now termed the magnetic constant) μ_0 and the absolute permittivity of free space (the electric constant) ε_0 are related by the electric field equation

$$\frac{1}{\sqrt{\mu_0\,\varepsilon_0}} = c_0 \qquad \ldots \qquad \ldots \qquad \ldots \qquad \ldots \quad (3)$$

where c_0 is the velocity of electromagnetic radiation in vacuo. The value of c_0 is

$$c_0 = 2 \cdot 997\ 925\ 0 \times 10^8 \pm 1 \cdot 0 \ \text{m/s}.$$

Thus the establishment of a value for one of these electromagnetic constants automatically defines the value of the other.

9–5 International (practical MKS) units: the MKSΩ system

A practical system of units based on experimental measurements of two electrical units, and related to the system of CGS electromagnetic units (see section 9–7), was first proposed in 1893 and adopted at the International Conference on Electrical Units and Standards in London in 1908. The two units concerned are:

(a) The international ampere A_{int}. The constant current that, when flowing through a solution of silver nitrate in water under specified conditions, deposits silver at a rate equal to its electrochemical equivalent in grammes per coulomb (g/C). The electrochemical equivalent of silver was measured to greater and greater degrees of accuracy; it was eventually standardised at 0·001 118 g/C. The present-day accepted value is 0·001 118 27 g/C.

(b) The international ohm Ω_{int}. The resistance at the temperature of melting ice of a column of mercury of mass 14·452 1 grammes, of length 106·300 centimetres, and of uniform cross section. The size of the cross-sectional area is extremely close to the one square millimetre used in an earlier definition. The international watt W_{int} is deduced from the theoretical relationship

$$P = I^2 R,$$

where P is the power in watts, I is the current in ampères and R is the resistance in ohms.

The size of the international ampère was so chosen in an attempt to make it equal to one-tenth of the size of the unit of current in the CGS electromagnetic system. It was soon recognised that one disadvantage of the method of definition is the need to change the sizes of the units as experimental procedures become more accurate. After standardising the numerical values of the units an even worse disadvantage became evident: the unit of power is not quite equal to the mechanical watt of one joule per second. It follows that the other international units (in particular the international joule) are not fundamentally related to the absolute mechanical MKS units. The system of international units was formally abandoned on 1 January 1948 at the 9th General Conference of Weights and Measures, which defined the relationships between the international and absolute units of voltage V (in volts V) and resistance, and therefore of current (from Ohm's law $V = IR$) and power ($= V^2/R$) and other quantities, as follows:

$$1 \text{ V}_{int} = 1·000 \text{ } 34 \text{ V}_{abs} \text{ exactly,}$$
$$1 \text{ } \Omega_{int} = 1·000 \text{ } 49 \text{ } \Omega_{abs} \text{ exactly;}$$
hence
$$1 \text{ A}_{int} = 0·999 \text{ } 850 \text{ A}_{abs} \text{ to six significant figures,}$$
$$1 \text{ W}_{int} = 1·000 \text{ } 19 \text{ W}_{abs} \text{ to six significant figures.}$$

It should be noted that the international system of units was never complete, having units for common current electrical and magnetic quantities only. Since it was formulated before the concept of rationalisation was considered the system is unrationalised. However, the limited number of units that were used apply to quantities occurring in equations almost all of which are identical in the two systems.

9–6 Absolute MKS units

These units comprise part of the International System (SI), sometimes referred to as the MKSA system (A = ampère). It is also called the Giorgi system because at the 1938 meeting of the International Electrotechnical Commission in Torquay, Professor Giorgi agreed to prepare a booklet (subsequently unpublished because of the outbreak of war in 1939) on the units of the MKSA (absolute) system. The question of rationalisation was originally left open; now only a rationalised system is employed. The four basic units involved are those of length (metre), mass (kilogramme), time (second), and current (ampère). The absolute permeability and permittivity appear explicitly as physical quantities (ie with size and units) in all relevant equations.

The definition of the ampère is the intensity of a constant current which, if maintained in two rectilinear parallel conductors of infinite length, of negligible circular cross section, and separated a distance one metre apart in a vacuum, would produce between these conductors a force equal to 2×10^{-7} newtons per metre of length. The value of the magnetic constant is

$$\mu_0 = 4\pi \times 10^{-7} \text{ henries per metre (H/m)}$$
$$= 1 \cdot 256\ 64 \times 10^{-6} \text{ H/m to six significant figures.}$$

(The use of the unit H/m saves the writing of more complicated units involving those of force and current.) The reason for the sizes of the quantities in these definitions arises from ensuring the connexion between the MKS and CGS electromagnetic units of current embodied in the basis of the older MKS system (see section 9–5).

(a) The size of μ_0

Using equation (1) for free-space conditions,

$$\frac{F}{l} \propto 2\mu_0 \frac{I_1 I_2}{x}.$$

1 SI unit of F (newton, N) $= 10^5$ CGS units (dyne),
1 SI unit of l, x (metre) $\quad = 10^2$ CGS units (centimetre),
1 SI unit of I (ampère) $\quad\ = 10^{-1}$ CGS electromagnetic units (abampère);

$$\mu_0 = 1 \text{ (CGS electromagnetic system).}$$

$$\therefore\quad \frac{10^{-5}}{10^{-2}} = \frac{(\mu_0/4\pi)(\text{SI rationalised})}{1\ (\text{CGS unrationalised})} \cdot \frac{10.10}{10^{-2}}$$

$$\mu_0 = 4\pi \times 10^{-7} \text{ SI rationalised units.}$$

(b) The definition of I

Substituting the value of μ_0 in the rationalised form of equation (1) for free-space conditions

$$\frac{F}{l} = \frac{2\mu_0}{4\pi} \frac{I_1 I_2}{x}$$

when $l = 1$ m, $x = 1$ m,

$$F = 2 \times 10^{-7} \text{ N}$$

occurs when

$$I_1 = I_2 = 1 \text{ A.}$$

(c) The size of the electric constant ε_0
This can be deduced using equation (3). It is

$$\varepsilon_0 = \frac{10^7}{4\pi(c)^2} \text{ farads per metre (F/m)}$$
$$= 8\cdot854\ 16 \times 10^{-12} \text{ F/m to six significant figures.}$$

(The use of the unit F/m saves the writing of more complicated units involving those of force and charge.) (c) is the numerical value of c_0 expressed in metres per second (see section 9–4), ie

$$(c) = c_0/(1 \text{ m/s}).$$

If an unrationalised absolute MKS system were required, the unrationalised form of equation (1) is used together with

$$\mu_0 = 10^{-7} \text{ H/m},$$
$$\varepsilon_0 = 10^7/(c)^2 \text{ F/m}.$$

9–7 CGS electromagnetic system

The units in this sytems are often referred to simply as electromagnetic units, CGS-emu, and are always unrationalised. Only three basic units are involved, those of length (centimetre, cm), mass (gramme, g) and time (second, s). The magnetic constant μ_0 is taken to have the value of unity, and is regarded as a numeric (ie having no units). Thus equation (1) for free-space conditions (unrationalised) becomes

$$\frac{F}{l} = \frac{2\,I_1\,I_2}{x},$$

and the definition of the CGS-emu of current is the intensity of a constant current which, if maintained in two rectilinear parallel conductors of infinite length, of negligible circular cross section, and separated by a distance of one centimetre in a vacuum, would produce between these conductors a force equal to two dynes per centimetre of length.

There are several points of importance.

(a) In those equations which should contain μ this quantity may be omitted altogether for free-space (and, practically, air) conditions: for other media μ must be included in the equations, but there is no distinction made between the relative and absolute values.

(b) The dimensions of I can be expressed in terms of [M], [L] and [T]. From equation (1) above the dimensions of I are clearly the same as those of the square root of force, ie $[M^{\frac{1}{2}} L^{\frac{1}{2}} T^{-1}]$.

(c) The name of the CGS-emu of current can be similarly obtained: it is $g^{\frac{1}{2}}\ cm^{\frac{1}{2}}/s$ or $dyne^{\frac{1}{2}}$. A convenient system of naming has been devised whereby each CGS-emu takes the name of the corresponding SI unit preceded by the prefix ab-. Eg the CGS-emu of current ($dyne^{\frac{1}{2}}$) is termed the abampère. Four

CGS-emus have been given special names, those of magnetic field strength (oersted), magnetic flux density (gauss), magnetic flux (maxwell) and magnetomotive force (gilbert).

(d) The size and units of ε_0 are calculated using equation (3):

$$\varepsilon_0 = 1/10^4 \, (c)^2 \, s^2/cm^2.$$

9–8 CGS 4-quantity electromagnetic system

The system is referred to as the CGSm system and is always unrationalised. In addition to the three mechanical units of the CGS-emu system the biot, Bi, is used as the unit of current, with a definition identical to that of the abampère. μ_0 also has a numerical value of unity, but its units are derived from the free-space unrationalised form of equation (1),

$$\frac{F}{l} = 2\mu_0 \frac{I_1 I_2}{x}$$

and are dynes per biot squared (dyn/Bi²). Using equation (3) the corresponding value of ε_0 is

$$\varepsilon_0 = 1/10^4 \, (c)^2 \, Bi^2 \, s^2/dyn \, cm^2.$$

All units in the CGSm and CGS-emu systems correspond to one another since

$$1 \, Bi = 1 \, abA \triangleq dyn^{\frac{1}{2}},$$

and the special names for the CGS-emus may be used for the corresponding CGSm units.

9–9 CGS electrostatic system.

The units in this system are often referred to simply as electrostatic units, CGS-esu, and are always unrationalised. Only three basic units are involved, those of length (centimetre), mass (gramme) and time (second). The electric constant ε_0 is taken to have the value of unity, and is regarded as a numeric. Thus equation (2) for free-space conditions (unrationalised) becomes

$$F = \frac{Q_1 Q_2}{x^2},$$

and the definition of the CGS-esu of charge is the value of that charge which, separated by a distance of one centimetre in a vacuum from another charge of equal size and nature, would produce between the charges a force equal to one dyne.

There are several points of importance.

(a) In those equations which should contain ε this quantity may be omitted altogether for free-space (and, practically, air) conditions: for other media ε

must be included in the equations, but there is no distinction between the relative and absolute values.

(b) The dimensions of Q can be expressed in terms of $[M]$, $[L]$ and $[T]$. From equation (2) above the dimensions of Q are clearly the same as those of the square root of force multiplied by those of distance, ie $[M^{\frac{1}{2}} L^{3/2} T^{-1}]$.

(c) The name of the CGS-esu of charge can be similarly obtained: it is $g^{\frac{1}{2}}$ cm$^{3/2}$/s or dyn$^{\frac{1}{2}}$ cm. A convenient system of naming has been devised whereby each CGS-esu takes the name of the corresponding SI unit preceded by the prefix stat-. Eg the CGS-esu of charge (dyn$^{\frac{1}{2}}$ cm) is termed the stat-coulomb. Unlike in the CGS-emu system (section 9–7(c)), none of the CGS-esus have been given special names.

(d) The size and units of μ_0 are calculated using equation (3):

$$\mu_0 = 1/10^4 \ (c)^2 \ \text{s}^2/\text{cm}^2.$$

9–10 CGS 4-quantity electrostatic system

The system is referred to as the CGSe system and is always unrationalised. In addition to the three mechanical units of the CGS-esu system the franklin, Fr, is used as the unit of charge, with a definition identical to that of the statcoulomb. ε_0 also has a numerical value of unity, but its units are derived from the free-space unrationalised form of equation (2),

$$F = \frac{Q_1 Q_2}{\varepsilon_0 \ x^2}$$

and are franklin squared per dyne centimetre squared (Fr^2/dyn cm^2). Using equation (3) the corresponding value of μ_0 is

$$\mu_0 = 1/10^4 \ (c)^2 \ \text{dyn s}^2/\text{Fr}^2.$$

All units in the CGSe and CGS-esu systems correspond to one another since

$$1 \ \text{Fr} \ \hat{=} \ 1 \ \text{statC} \ \hat{=} \ \text{dyn}^{\frac{1}{2}} \ \text{cm}.$$

9–11 CGS mixed system

The system is also called the Gaussian or symmetric system and is unrational-ised. It employs the three basic mechanical units (centimetre, gramme and second), and uses units from the CGS-emu system for magnetic quantities and those from the CGS-esu for electrical quantities. Because of this mixture the value (c) appears explicitly in some equations that involve quantities derived from both systems. $\mu_0 = 1$; $\varepsilon_0 = 1$. Table 10 lists the quantities employed from each system, with their symbols.

Table 10: The quantities of the CGS mixed system

CGS-emu quantities		CGS-esu quantities	
		current	I
		charge	Q
magnetic field strength	H	electric field strength	E
magnetic potential and pd	U	electric potential and pd	V
magnetomotive force	F	electromotive force	E
magnetic induction	B	displacement	D
magnetic flux	Φ	electric flux	Ψ
self and mutual inductance	L, M	capacitance	C
permeability	μ	permittivity	ε
magnetic polarisation	J	electric polarisation	P
magnetic moment	M	electric dipole moment	p
reluctance	R	resistance	R
permeance	Λ	conductance	G
		resistivity	ρ
		conductivity	σ

If desired, the Gaussian mixed system may employ a combination of units from the CGSm and CGSe systems (instead of from the CGS-emu and CGS-esu systems).

9–12 Heaviside-Lorentz system

This is the rationalised form of the Gaussian system, based on the free-space rationalised equations

$$\frac{F}{l} = \frac{\mu_0}{2\pi} \frac{I_1 I_2}{x}$$

and

$$F = \frac{Q_1 Q_2}{4\pi \, \varepsilon_0 \, x^2}.$$

It should be noted that the equation originally used by Heaviside and Lorentz to obtain the units of magnetic quantities was that for the force F between two 'magnetic poles' m_1, m_2, at a separation x,

$$F = \frac{m_1 m_2}{4\pi \, \mu_0 \, x^2}.$$

Current was obtained by application of the Ampère-Laplace theorem.
The units of the more important quantities have the same sizes as those in the Gaussian system, except that the numerical sizes of the magnetic and electric constants are

$$\mu_0 = 4\pi \text{ and}$$
$$\varepsilon_0 = 1/4\pi :$$

the unit of magnetic pole strength is 4π times the size of the unrationalised unit, and the rationalised defining equations (given in the entries in part 2, Quantities) must be used.

9–13 Other metric systems

(a) MTS system
An electrical system based on the MTS mechanical system can be designed having electrical units identical with those of the SI (generally rationalised), but with a suitable value given to the magnetic constant, and hence to the electric constant. Using equation (1) for free-space conditions,

$$\frac{F}{l} \propto 2\mu_0 \frac{I_1 I_2}{x}.$$

1 MTS unit of F (sthène) = 10^3 SI units;
the unit of length (metre) is common to both systems;
$\mu_0 = 4\pi \times 10^{-7}$ H/m (SI rationalised).

$$\therefore \quad 10^{-3} = \frac{\mu_0 \text{ (MTS rationalised)}}{4\pi \times 10^{-7}}.$$

$$\mu_0 = 4\pi \times 10^{-10} \text{ H/m (MTS rationalised)}.$$

From equation (3)

$$\varepsilon_0 = 10^{10}/4\pi \, (c)^2 \text{ F/m (MTS rationalised)}.$$

(b) Mie system
This is developed in a similar manner to the development of the MTS system.
1 Mie unit of $F = 10^2$ SI units;
1 Mie unit of $l, x = 10^{-2}$ SI units;
$\mu_0 = 4\pi \times 10^{-7}$ H/m (SI rationalised).

$$\therefore \quad \frac{10^{-2}}{10^2} = \frac{\mu_0 \text{ (Mie rationalised)}}{4\pi \times 10^{-7}} \cdot \frac{1.1}{10^2}.$$

$$\mu_0 = 4\pi \times 10^{-9} \text{ H/m (Mie rationalised)}.$$

From equation (3)

$$\varepsilon_0 = 10^9/4\pi \, (c)^2 \text{ F/m (Mie rationalised)}.$$

(c) MkgfS system
This follows as in the previous examples.
1 MkgfS unit of F (kilogramme-force) = (g) SI units;
the unit of length (metre) is common to both systems;
$\mu_0 = 4\pi \times 10^{-7}$ H/m (SI rationalised).
(g) is the numerical value for the standard acceleration of free fall g_n in metres per second squared. Ie

where
$$(g) = g_n/(1 \text{ m/s}^2),$$
$$g_n = 9 \cdot 806\,65 \text{ m/s}^2.$$
$$\frac{1}{(g)} = \frac{\mu_0 \text{ (MkgfS rationalised)}}{4\pi \times 10^{-7}}.$$
$$\mu_0 = 4\pi \times 10^{-7}/(g) \text{ H/m (MkgfS rationalised)}.$$

From equation (3)

$$\varepsilon_0 = (g) \times 10^7/4\pi \, (c)^2 \text{ F/m (MkgfS rationalised)}.$$

A system of this kind has never been formulated for use.

9–14 Imperial systems

There has never been a desire to formulate electrical units based on any of the imperial systems of units, although these could be done as analogues to the CGS-emu and CGS-esu systems (with μ_0 or ε_0 equal to unity), or even to the SI (with μ_0 and ε_0 possessing suitable values). The basis for the FPS and FSS unrationalised systems (as analogues of the CGS-emu and CGS-esu systems) is given in table 11.

Table 11: FPS- emu and -esu and FSS-emu and -esu units

	emu	esu
basic equation for free-space conditions given value of one free-space constant	(1) $\dfrac{F}{l} = 2\mu_0 \dfrac{I_1 I_2}{x}$ $\mu_0 = 1$	(2) $F = \dfrac{Q_1 Q_2}{\varepsilon_0 x^2}$ $\varepsilon_0 = 1$
(a) FPS systems units of quantities in basic equation $\{$ $\therefore$ derived unit $\therefore$ from $I = dQ/dt$, unit of $\therefore$ value of 2nd free-space constant (from (3))*	F: poundal (pdl) l, x: foot (ft) I: pdl$^{\frac{1}{2}}$ Q: pdl$^{\frac{1}{2}}$ s $\varepsilon_0 = k/(c)^2$ s^2/ft^2	F: poundal (pdl) x: foot (ft) Q: pdl$^{\frac{1}{2}}$ ft I: pdl$^{\frac{1}{2}}$ ft/s $\mu_0 = k/(c)^2$ s^2/ft^2
(b) FSS systems units of quantities in basic equation $\{$ $\therefore$ derived unit $\therefore$ from $I = dQ/dt$, unit of $\therefore$ value of 2nd free-space constant (from (3))*	F: pound-force (lbf) l, x: foot I: lbf$^{\frac{1}{2}}$ Q: lbf$^{\frac{1}{2}}$ s $\varepsilon_0 = k/(c)^2$ s^2/ft^2	F; pound-force (lbf) x: foot Q: lbf$^{\frac{1}{2}}$ ft I: lbf$^{\frac{1}{2}}$ ft/s $\mu_0 = k/(c)^2$ s^2/ft^2

*Since   1 ft = 30·48 cm,
   1 m^2/ft^2 = 30·48^2 cm^2/ft^2 ÷ 10^4 cm^2/m^2
         = 0·092 903 04 exactly.
Thus k = 0·092 903 04.

The units of the common electrical and magnetic quantities in these two systems can be derived from the mechanical-equivalent units of the CGS-emu and CGS-esu systems (table 13) by replacing centimetre by foot, dyne by poundal or pound-force, and erg by foot poundal or foot pound-force.

9–15 The values of the magnetic and electric constants in various systems of units: a summary

Table 12 contains rationalised values; for unrationalised values, the factor 4π should be omitted in each case.

Table 12: Values of the magnetic and electric constants

system	magnetic constant		electric constant	
SI; MKSA; MKSΩ	$4\pi \times 10^{-7}$	H/m	$10^7/4\pi(c)^2$	F/m
CGS-emu	4π		$10^{-4}/4\pi(c)^2$	s^2/cm^2
CGSm	4π	dyn/Bi^2	$10^{-4}/4\pi(c)^2$	$Bi^2\ s^2/dyn\ cm^2$
CGS-esu	$4\pi/10^4(c)^2$	s^2/cm^2	$1/4\pi$	
CGSe	$4\pi/10^4(c)^2$	$dyn\ s^2/Fr^2$	$1/4\pi$	$Fr^2/dyn\ cm^2$
Gaussian (symmetric) ⎫ Heaviside-Lorentz ⎭	4π		$1/4\pi$	
Gaussian*	4π	dyn/Bi^2	$1/4\pi$	$Fr^2/dyn\ cm^2$
MTS	$4\pi \times 10^{-10}$	H/m	$10^{10}/4\pi(c)^2$	F/m
Mie	$4\pi \times 10^{-9}$	H/m	$10^9/4\pi(c)^2$	F/m
MkgfS	$4\pi/10^7(g)$	H/m	$10^7(g)/4\pi(c)^2$	F/m
FPS-emu; FSS-emu	4π		$k/4\pi(c)^2$	s^2/ft^2
FPS-esu; FSS-esu	$4\pi k/(c)^2$	s^2/ft^2	$1/4\pi$	

* Based on CGSm and CGSe systems rather than CGS-emu and CGS-esu systems

$(c) = 2 \cdot 997\ 925\ 0 \times 10^8$ (see section 9–6(c));
$(g) = 9 \cdot 806\ 65$ (see section 9–13(c));
$k = 0 \cdot 092\ 903\ 04$ (see footnote to table 11).

9–16 Named units in the metric systems

Table 13 lists all the fundamental and derived units of the SI; CGSm, CGS-emu, CGSe, CGS-esu; Gaussian and Heaviside-Lorentz systems that have special names. Other derived units are named according to the rules given in section 3(b). Because of the extent of the coverage of the table, a number of units are included with names that are combination forms.

For the other systems:

(a) The MKSA and MKSΩ electrical units are those of the SI.

(b) By extension, the MkgfS, MTS and MTS-thermal electrical units are those of the SI.

(c) By extension, the Mie electrical units are those of the SI, except in the cases noted.

Table 13: Electrical units

quantity	SI	CGSm	CGS-emu			CGSe	CGS-esu	
			1	2	3		4	3
capacitance	F	$Bi^2 s^2/$erg	—	abF	s^2/cm	$Fr^2/$erg	statF	cm*
charge	C	Bi s	—	abC	$dyn^{\frac{1}{2}} s$	Fr	statC	$dyn^{\frac{1}{2}}$ cm*
conductance[5]	S	$Bi^2 s/$erg	—	abS	s/cm	$Fr^2/$erg s	statS	cm/s*
current	A	Bi	—	abA	$dyn^{\frac{1}{2}}$	Fr/s	statA	$dyn^{\frac{1}{2}}$ cm/s*
field strength, magnetic[6]	A/m	Bi/cm	Oe	abA/cm	$dyn^{\frac{1}{2}}/$cm*	Fr/cm s	statA/cm	$dyn^{\frac{1}{2}}/s$
flux, magnetic	Wb	erg/Bi	Mx	abWb	$dyn^{\frac{1}{2}}$ cm*	erg s/Fr	statWb	$dyn^{\frac{1}{2}} s$
flux density, magnetic[7]	T	erg/$Bi cm^2$	G	abT	$dyn^{\frac{1}{2}}/$cm*	erg s/$Fr cm^2$	statT	$dyn^{\frac{1}{2}} s/cm^2$
inductance	H	erg/Bi^2	—	abH	cm*	erg s^2/Fr^2	statH	s^2/cm
magnetomotive force[8]	A(-T)	Bi	Gb	abA	$dyn^{\frac{1}{2}}$*	Fr/s	statA	$dyn^{\frac{1}{2}}$ cm/s
power, apparent	VA[9]	erg/s	—	—	erg/s*	erg/s	—	erg/s*
power, reactive	var[9]	erg/s	—	—	erg/s*	erg/s	—	erg/s*
resistance	Ω	erg/Bi^2 s	—	abΩ	cm/s	erg s/Fr^2	statΩ	s/cm*
voltage	V	erg/Bi s	—	abV	$dyn^{\frac{1}{2}}$ cm/s	erg/Fr	statV	$dyn^{\frac{1}{2}}$*

Unit symbols employed:

A	ampère	dyn	dyne	Gb	gilbert	Oe	oersted	V	volt
Bi	biot	F	farad	H	henry	S	siemens	Wb	weber
C	coulomb	Fr	franklin	m	metre	s	second	Ω	ohm
cm	centimetre	G	gauss	Mx	maxwell	T	tesla		

* Gaussian unit. The other corresponding CGS-emu (or CGS-esu) may be employed. Alternatively, the corresponding CGSm (or CGSe) unit may be employed. These are also the units of the Heaviside-Lrentz system.

1 Special name (where it exists).

2 Name based on the principle of placing the prefix ab- in front of the corresponding SI unit.

3 Mechanical equivalent.

4 Name based on the principle of placing the prefix stat- in front of the corresponding SI unit.

5 The SI unit name is not recognised internationally, although it is often used. The name per ohm is employed.

6 The unit in the Mie system is A/cm.

7 The unit in the Mie system is Wb/cm^2.

8 The SI unit, ampère-turn, is now normally just called ampère.

9 = watt.

(d) The SI is an extension of the MKS system; there are no electrical units in the latter.

(e) The various CGS electrical systems are extensions of the CGS, CGS-thermal and CgfS systems; there are no electrical units in the three last-named systems.

(f) There are no electrical units in the mm-mg-s, dm-kg-ds and MGS systems, although they can be derived if necessary. The units would be decimal multiples or submultiples of the SI and CGS electrical units.

(g) None of the possible units of the FPS, FPS-thermal, FSS and FlbfS systems have been tabulated. They can be derived as in section 9–14, but are never employed. This also applies to the Stroud and ft-gr-s systems.

10 Natural systems of units

The Hartree system is often employed in the study of atomic structure, and the Quantum-electrodynamics system in the study of radiations (eg β- and γ-rays). Thus in neither system is it necessary to provide a complete set of units. The most important units of the two systems have been collected together in table 14. For two of the fundamental quantities (h and e) the sixth significant figure in the measured value is in doubt; for a third (m) this is true even for the fifth significant figure. The values in the table have therefore been given to four significant figures.

11 The Ludovici system of units

This system has been proposed with the suggested virtue that it is built upon the most basic units available. As the table shows, a great drawback is in the extremely small sizes of both its fundamental and derived units. Because of the uncertainty in the fourth significant figure in the measured value of G, all the tabulated values are given to three significant figures. The sizes of the derived units have been obtained from their dimensional forms (see Appendix 2), these being obtained using the fundamental formulae of electricity and magnetism (as given in section 9–2) and Newton's law of universal gravitation,

$$F = \frac{G\, m_1\, m_2}{x^2}.$$

(F is the force between masses m_1, m_2 at a separation x.) Because the magnetic and electric constants are related through the electric field equation (see section 9–4), the dimensional forms involving the velocity of propagation of electromagnetic waves in vacuo, c, are most useful.

Table 14: Units of the natural systems

quantity	Hartree system		Quantum-electrodynamics system		ratio[1]
	rationalised definition	size in SI units[2]	rationalised definition	size in SI units[2]	
FUNDAMENTAL QUANTITIES					
mass	m (note 3)	$9{\cdot}108 \times 10^{-31}$ kg	m (note 3)	$9{\cdot}108 \times 10^{-31}$ kg	1
action[4]	$\hbar$ $[= h/2\pi]$ (note 5)	$1{\cdot}054 \times 10^{-34}$ J s	$\hbar$ $[= h/2\pi]$ (note 5)	$1{\cdot}054 \times 10^{-34}$ J s	1
charge	e (note 6)	$1{\cdot}602 \times 10^{-19}$ C	—	—	$K^{-\frac{1}{2}}$
velocity	—	—	c (note 7)	$2{\cdot}998 \times 10^{8}$ m/s	K^{-1}
electric constant	ε_o $[= 10^7/4\pi c^2]$	$8{\cdot}854 \times 10^{-12}$ F/m	ε_o $[= 10^7/4\pi c^2]$	$8{\cdot}854 \times 10^{-12}$ F/m	1
DERIVED QUANTITIES					
charge			$(4\pi\varepsilon_o \hbar c)^{\frac{1}{2}}$	$1{\cdot}875 \times 10^{-18}$ C	$K^{-\frac{1}{2}}$
velocity	$a_o/\tau_o = e^2/4\pi\varepsilon_o \hbar$	$2{\cdot}188 \times 10^{6}$ m/s	—	—	K^{-1}
length	$a_o = 4\pi\varepsilon_o \hbar^2/m\,e^2$ (note 8)	$5{\cdot}292 \times 10^{-11}$ m	$\lambda_{ce} = \hbar/m\,c$ (note 9)	$3{\cdot}862 \times 10^{-13}$ m	K
time	$\tau_o = 16\pi^2\varepsilon_o^2 \hbar^3/m\,e^4$ (note 10)	$2{\cdot}419 \times 10^{-17}$ s	$\lambda_{ce}/c = \hbar/m\,c^2$	$1{\cdot}288 \times 10^{-21}$ s	K^2
energy	$e^2/a_o^2 = m\,e^4/16\pi^2\varepsilon_o^2 \hbar^2$ (note 11)	$4{\cdot}851 \times 10^{-18}$ J	$m\,c^2$ (note 12)	$8{\cdot}186 \times 10^{-14}$ J	K^{-2}

1 Size of the Hartree system unit divided by the size of the Quantum-electrodynamics system unit, where

$$K = 4\pi\,\varepsilon_o\,\hbar\,c/e^2$$
$$= 10^7\,\hbar/e^2 c$$
$$= 137{\cdot}0.$$

2 Unit symbols employed:

C coulomb	m metre	kg kilogramme
F farad	J joule	s second

3 The rest mass of the electron.
4 = angular momentum in kg m²/s.
5 h is Planck's constant.
6 The charge on the electron.
7 The velocity of propagation of electromagnetic waves in a vacuum.
8 The radius of the first Bohr orbit (for infinite mass).
9 = $\lambda_{ce}/2\pi$; λ_{ce} is the Compton wavelength of the electron.
10 The period of the first Bohr orbit.
11 The unit is sometimes called the hartree.
12 The energy-equivalent rest mass of the electron

Table 15: Important quantities of the Ludovici system

quantity	dimensional form								size in SI units[1]
	G	e	μ	ε	G	e	μ	c	
FUNDAMENTAL QUANTITIES									
constant of universal gravitation	1	0	0	0	1	0	0	0	$6{\cdot}67 \times 10^{-11}$ N m²/kg²
charge on the electron	0	1	0	0	0	1	0	0	$1{\cdot}60 \times 10^{-19}$ C
magnetic constant[2]	0	0	1	0	0	0	1	0	$1{\cdot}26 \times 10^{-6}$ H/m
electric constant[3]	0	0	0	1	0	0	−2	−1	$8{\cdot}85 \times 10^{-12}$ F/m
DERIVED QUANTITIES									
length	½	1	1	½	½	1	½	−1	$4{\cdot}89 \times 10^{-36}$ m
mass	−½	1	0	−½	−½	1	½	1	$6{\cdot}59 \times 10^{-9}$ kg
time	½	1	$\frac{3}{2}$	1	½	1	½	−2	$1{\cdot}63 \times 10^{-44}$ s

1 Unit symbols employed:

C	coulomb	m	metre
F	farad	N	newton
H	henry	s	second
kg	kilogramme		

2 $4\pi \times 10^{-7}$ H/m.

3 $10^7/4\pi(c)^2$ F/m. (c) is the numerical value of c, the velocity of electromagnetic radiation in vacuo.

Appendix 2

DIMENSIONS

1 The meaning of dimensions

When the quantities length (l) and time (t) are considered it is clear that they are totally independent concepts, having nothing in common with one another. Similarly, the quantities area (A) and frequency (f) are independent concepts; they are, however, conceptually related directly to length and time respectively. The relations are given by the defining equations

$$f = \frac{1}{t} \qquad \ldots \qquad \ldots \qquad \ldots \qquad \ldots \quad (1)$$

and
$$A \propto l_1 l_2 \qquad \ldots \qquad \ldots \qquad \ldots \qquad \ldots \quad (2)$$

(For equation (2) the value of the constant of proportionality depends on the nature of the area. Eg for a triangle, with l_1 representing the base length and l_2 the height, the constant has the value $\frac{1}{2}$; for a circle l_1 and l_2 are both equal to the radius, and the constant has the value π.) The quantity velocity (v) is conceptually related to both length and time since it is defined (for uniform velocity) by the equation

$$v = \frac{l}{t} \qquad \ldots \qquad \ldots \qquad \ldots \qquad \ldots \quad (3)$$

Although it would not be valuable to define velocity in terms of area and frequency there is quite clearly a formular relationship between them.

The physical nature or concept of a quantity is termed its dimensional form, this being a complex combining the dimensions of quantities with simpler dimensional forms. Thus length and area both have dimensional forms that contain exactly the same kind of dimensions, as do both time and frequency, although the nature of the dimensions of the latter pair are quite different from those of the former pair. The dimensional form of velocity will show dimensions that relate it to both length (or area) and time (or frequency).

2 Fundamental dimensions

Just as certain quantities (mass, length, time, etc) are regarded as fundamental and possess unrelated fundamental units (see Appendix 1, section 3(a)), so certain quantities are given unrelated fundamental dimensions. The choice of these quantities is arbitrary and need have no connexion with the choice of quantities with fundamental units, provided that there is no equation that relates them. This is obvious because, if such an equation exists, the dimensions of one of the quantities is automatically derived from the others. Hence the fundamental dimensional quantities cannot include length, time and velocity, these three being related to one another through equation (3).

The choice of the quantities with fundamental dimensions is made bearing the following points in mind:

(a) the dimensions of other quantities which are themselves not fundamental can be easily derived;

(b) if possible, they appear in dimensional forms in a simple manner (eg raised to powers that are integral and of small size).

To fulfil the first criterion it is clearly easiest in practice to give the quantities with fundamental units fundamental dimensions too: defining laws are often directly, and otherwise always indirectly, based on the quantities with fundamental units. It almost certainly follows that the second criterion is also fulfilled.

If it is thought necessary to decide between time and frequency (which are simply related by inverse equality), the latter will be found to possess the marginal advantage of more often being raised to positive powers than the former. This does not really outweigh the advantage that so many quantities are defined in terms that include the word 'time' (eg velocity, acceleration, power, impulse, charge). The choice of area rather than length is unsatisfactory, since in all formulae in which length appears explicitly it would have to be represented by the square root of area.

The dimensions of quantities are represented by convenient letters, and the dimensional form is usually enclosed in square brackets. The quantity itself is also enclosed in square brackets when it is made the subject of a dimensional equation, and it can therefore be distinguished from the size of the quantity. The individual members of a dimensional form are written consistently in the same order. In this work the order of dimensions has been chosen to reflect roughly how often they are employed (mechanical dimensions precede those of non-mechanical quantities), and so that those more often raised to positive powers precede those more often raised to negative powers.

The number of fundamental dimensions required can only be determined by a systematic examination of every empirical law, defining equation and derived relationship. The choice of length alone would only allow dimensional forms for area, volume, second moment and curvature; the addition of time allows for frequency, angular acceleration, velocity, acceleration, gravitational potential, etc. Many quantities could not, with these two alone, be given dimensional forms, quantities such as force and voltage. It appears probable that the number of fundamentals required in any given branch of physical science is equal to the number of quantities with fundamental units used in that branch. To cover physical science completely it seems at present that the number required is six (or seven, if molar value is included: see the international system of units, SI, table 1).

3 Types of quantities

There are four classes of dimensional and non-dimensional quantities.

Appendix 2

(a) Dimensioned variables
Most variable quantities fall into this class. It includes both quantities the dimensions of which are regarded as fundamental (eg length and time) and those that are derived (eg area and velocity).

(b) Dimensionless variables
Some variables are so defined that the combination of dimensions of the component quantities that comprise them cancel one another out. They are either:
  (i) ratios, eg the coefficient of statical friction and relative density; or
  (ii) numerics, ie quantities that combine more than two components in such a way, often with deliberate intention, that the dimensions of the components cancel. An example of a numeric is the Reynolds number.

(c) Dimensioned (ie physical) constants
This class includes the velocity of electromagnetic radiation in vacuo and the constant of universal gravitation.

(d) Dimensionless (ie mathematical) constants
This class includes all types of numbers (integers, complex, transcendental, etc). It follows that the dimensional form of a derived quantity does not contain reference to any dimensionless coefficients included in the defining equation. Thus the dimensional form of kinetic energy, $\frac{1}{2}mv^2$, (m = mass, v = velocity) is that given to the product mv^2; it is identical to that given to the product mgh (g = acceleration of free fall, h = height above the datum level) which defines potential energy, and to any other quantity that is a form of energy.

4 The dimensions of mechanical (and acoustical) quantities

4-1 [M], [L] and [T]
An analysis of all mechanical and acoustical formulae shows that only three fundamentals are required. For practically every purpose those corresponding to the fundamental quantities of mechanics are chosen, ie mass [M], length [L] and time [T]. It is usual for the symbol given to a fundamental dimensional form to be the capital letter equivalent of that of the quantity it represents.
To explain the method of construction of derived dimensional forms, and to provide a reference list of the dimensions of the more important mechanical quantities (including those required for the analyses given in section 9), a table of dimensions has been constructed. The dimensions of each quantity simply combines the dimensions of the members of its definition according to the defining equation. It will be noted that a given dimensional form is not necessarily unique to one quantity (see energy and torque; impulse and momentum).

Table 16: The dimensions of common mechanical quantities

quantity and symbol		defining formula*	dimensions M L T
acceleration	a	$a = dv/dt$	0 1 −2
action	L	$L = \int E dt$	1 2 −1
area	A	$A \propto l_1 l_2$	0 2 0
density	ρ	$\rho = m/V$	1 −3 0
elastic modulus	E	$E = \dfrac{F}{A} \Big/ \dfrac{\Delta l}{l}$	1 −1 −2
energy	E	$E = Fs$	1 2 −2
force	F	$F = ma$	1 1 −2
frequency	f	$f = 1/T$	0 0 −1
impulse	I	$I = \int F\,dt$	1 1 −1
length	l	—	0 1 0
mass	m	—	1 0 0
moment of inertia	I	$I = \Sigma m\,r^2$	1 2 0
momentum	p	$p = mv$	1 1 −1
power	P	$P = E/t$	1 2 −3
pressure	p	$p = F/A$	1 −1 −2
reciprocal length	$1/l$	—	0 −1 0
surface tension	γ	$\gamma = F/l$	1 0 −2
time	t	—	0 0 1
torque	M	$M = F \times r$	1 2 −2
velocity	v	$v = ds/dt$	0 1 −1
viscosity (dynamic)	η	$\eta = \tau \Big/ \dfrac{dv}{dz}$	1 −1 −1
viscosity, kinematic	ν	$\nu = \eta/\rho$	0 2 −1
volume	V	$V \propto l_1 l_2 l_3$	0 3 0

* Quantity symbols for quantities that do not appear in the table:

r = radius vector (l)
s = displacement (l)
T = periodic time (t)
z = distance across flow (l)
τ = tangential stress (F/A)

4–2 [F], [L] and [T]

For certain limited purposes it has been suggested that it is more convenient to take the dimensions of force as fundamental instead of those of mass, together with those of length and time. Since

$$[F] = [M\ L\ T^{-2}],$$

the dimensional forms of the quantities tabulated (and others) can be obtained by replacing [M] by [F L^{-1} T^2]: eg that for power becomes [F L T^{-1}].
Only the [M], [L], [T] system is given in the entries in part 2 (Quantities).

5 The dimensions of thermal quantities

5–1 [M], [L], [T], [Q] and [Θ]

The dimensions assigned to a thermal quantity will depend on the physical nature assumed to be associated with quantity of heat (Q) and temperature (θ). If these quantities are assumed to possess dimensional characters of their own, or alternatively if they are considered to be basically different from corresponding mechanical quantities, they must be added to the three quantities length, mass and time. Thus the dimensions of specific heat capacity (mass basis) (c) are deduced from the defining equation

$$c = \frac{1}{m} \frac{dQ}{d\theta},$$

and
$$[c] = [M^{-1} \, Q \, \Theta^{-1}].$$

The particular advantage of this arrangement is that it provides for the maximum number of dimensions, five.

5–2 [M], [L] and [T]

It may alternatively be assumed that both Q and θ are identical to corresponding mechanical quantities:

  (a) Heat is a form of energy, and its dimensions are therefore [M L² T⁻²].
  (b) The kinetic theory suggests that temperature has the dimensions of velocity squared, [L² T⁻²].

The dimensions of specific heat capacity are, in this system, given by

$$[c] = \frac{1}{[M]} \frac{[M \, L^2 \, T^{-2}]}{[L^2 \, T^{-2}]}.$$

[c] is therefore dimensionless, this idea being strengthened by an alternative definition used when heat energy was measured in calories: the ratio of the quantity of heat required to raise the temperature of a given mass of a substance by a certain amount, divided by the quantity of heat required to raise the temperature of an identical mass of water by the same amount. (The present-day definition of specific heat capacity gives numerical values larger than those resulting from the former definition by the ratio of the calorie per gramme to the joule per kilogramme, 4 186·8.)

The particular advantage of this arrangement is that it provides for a system that is a natural extension of the system used in mechanics.

5–3 [M], [L], [T] and [Q] or [Θ]

Either of the two assumptions given in section 5–2 may be accepted and the other rejected. This will lead to two four-dimensional systems. Specific heat capacity in each of these systems has the following dimensions:

(a)
$$[c] = \frac{1}{[M]} \frac{[Q]}{[L^2\ T^{-2}]}$$
$$= [M^{-1}\ L^{-2}\ T^2\ Q];\ \text{or}$$

(b)
$$[c] = \frac{1}{[M]} \frac{[M\ L^2\ T^{-2}]}{[\Theta]}$$
$$= [L^2\ T^{-2}\ \Theta^{-1}].$$

Two other four-dimensional systems are possible if the concept of a dimensionless specific heat capacity is combined with the dimensions of c given in section 5–1, $[M^{-1}\ Q\ \Theta^{-1}]$. It will be clear that:
(a) retaining $[Q]$, the dimensions of θ are $[M^{-1}\ Q]$; or
(b) retaining $[\Theta]$, the dimensions of Q are $[M\ \Theta]$.
This is most clearly illustrated by the entries in table 17, giving the dimensional forms of the more important thermal quantities in each of the six systems given above. (The two systems described last of all are labelled MLTQ(2) and MLT Θ(2) to distinguish them from the first two described in this section, MLTQ(1) and MLT Θ(1).)
It will be seen from the table that simply because one quantity can have the same dimensions in two or more given systems, this does not follow for every quantity (see systems MLTQ Θ, MLTQ(1) and MLTQ(2) in respect of heat energy and temperature). In the entries in part 2 (Quantities), only the systems MLTQ Θ and MLT are given.

6 The dimensions of photometric quantities

The system of dimensions assigned to photometric quantities depends on the assumed nature of luminous flux. If flux is assumed to possess a dimensional nature of its own the four fundamentals mass, length, time and flux are required: flux is represented by [F]. Alternatively, luminous flux can be considered as identical to the corresponding mechanical quantity, power, and assigned the dimensions $[M\ L^2\ T^{-3}]$. Both systems are given in the entries in part 2 (Quantities). Flux is preferred to luminous intensity (the defined fundamental quantity of the SI) because of its more basic nature.

7 The dimensions of electrical and magnetic quantities

The choice of a fourth fundamental dimensional quantity in addition to mass, length and time may be made from the following list of those most commonly employed:
(a) magnetic permeability μ
(b) electrostatic permittivity ε
(c) electric charge Q
(d) electric current I
(e) electric resistance R

Table 17: The dimensions of common thermal quantities

quantity and symbol	defining equation*	M L T Q Θ					M L T			M L T Q (1)				M L T Q (2)				M L T Θ (1)				M L T Θ (2)			
		M	L	T	Q	Θ	M	L	T	M	L	T	Q	M	L	T	Q	M	L	T	Θ	M	L	T	Θ
heat energy, Q	—	0	0	0	1	0	1	2	-2	0	0	0	1	0	0	0	1	1	2	-2	0	1	0	0	1
temperature, θ	—	0	0	0	0	1	0	2	-2	0	2	-2	0	-1	0	0	1	0	0	0	1	0	0	0	1
specific heat capacity (mass basis), c	$c = \dfrac{1}{m}\dfrac{dQ}{d\theta}$	-1	0	0	1	-1	0	0	0	-1	-2	2	1	0	0	0	0	0	2	-2	-1	0	0	0	0
conductivity, thermal, λ	$\dfrac{dQ}{dt} = -\lambda A \dfrac{d\theta}{dx}$	0	-1	-1	1	-1	1	-1	-1	0	-3	1	1	1	-1	-1	0	1	1	-3	-1	1	-1	-1	0
diffusivity, thermal, α	$\alpha = \lambda/\rho c$	0	2	-1	0	0	0	2	-1	0	2	-1	0	0	2	-1	0	0	2	-1	0	0	2	-1	0
entropy, S	$dS = dQ/T$	0	0	0	1	-1	1	0	0	0	-2	2	1	1	0	0	0	1	2	-2	-1	1	0	0	0
heat capacity, C	$C = dQ/d\theta$	0	0	0	1	-1	1	0	0	0	-2	2	1	1	0	0	0	1	2	-2	-1	1	0	0	0
latent heat, specific, l	$l = Q/m$	-1	0	0	1	0	0	2	-2	-1	0	0	1	-1	0	0	1	0	2	-2	0	0	0	0	1
resistivity, thermal, ρ	$\rho = 1/\lambda$	0	1	1	-1	1	-1	1	1	0	3	-1	-1	-1	1	1	0	-1	-1	3	1	-1	1	1	0
temperature coefficient, α	$\alpha = \Delta X/X\Delta\theta$	0	0	0	0	-1	0	-2	2	0	-2	2	0	1	0	0	-1	0	0	0	-1	0	0	0	-1

* Quantity symbols for quantities that do not appear in the table:

A = area
m = mass
t = time
T = absolute temperature
x = distance
X = any quantity that varies with temperature
ρ = density

The decision is made with the following considerations in mind.

(i) the use of permeability is only significant in quantities that are directly related to magnetic measurements.

(ii) The use of permittivity is only significant in quantities that are directly related to electrostatic measurements.

(iii) The use of permeability or permittivity yields powers of the dimensional fundamentals that are in general fractional, and can therefore be awkward to operate with.

(iv) The choice of Q, I or R is somewhat arbitrary, since the three quantities are related through the simple defining equations

$$I = \frac{dQ}{dt}$$

and
$$P = I^2 R$$

(t = time, P = power). Current is a better choice than charge because although the latter has a more fundamental nature, the former is one of the fundamental quantities of the SI. Current also appears in far more defining equations than does charge. It is also a better choice than resistance for the same reason of being more commonly employed.

Two systems employing only the dimensions of mass, length and time can be formulated, assuming that the permeability or permittivity is dimensionless (ie regarding one of these quantities as relative). The two systems are different from one another because both μ and ε cannot be simultaneously dimensionless, being related through the electric field equation (see Appendix 1, section 9–4):

$$\frac{1}{\sqrt{\mu_0 \, \varepsilon_0}} = c_0$$

(c_0 is the velocity of electromagnetic radiation in vacuo). It cannot really be recommended that either μ or ε is regarded as dimensionless.

The systems MLTI, MLTμ and MLTε only are given in the entries in part 2 (Quantities).

8 Dimensional forms involving vector lengths

It has been proposed that although length is itself a fundamental quantity, it may be split up into three component vector lengths parallel to the axes of an applied three-dimensional rectangular Cartesian system of coordinates. This makes clear the vector nature of a quantity from its dimensional form, although it also causes vector components of length to appear in the dimensional forms of those scalar quantities that are defined from the scalar product of vectors, quantities such as volume (proportional to area multiplied by length) and energy (force multiplied by distance in the direction of the force). The dimensions of the vector components of length may be represented by [X], [Y] and [Z], and the orientation of the coordinate system must be explicitly stated. There are three dimensional forms for each quantity, by cyclicly mutating the three vector components of length.

Table 18 gives the dimensional forms of all the mechanical quantities also given in table 16. It may of course be extended to thermal, photometric, electrical and magnetic quantities, although its value in these branches of physics is likely to be rather limited.

Table 18: The vector dimensions of common mechanical quantities

| quantity | dimensions | | | | | orientation of the coordinates: quantity in the [X]-direction |
	M	X	Y	Z	T	
acceleration	0	1	0	0	-2	motion
action	1	2	0	0	-1	applied force
area	0	0	1	1	0	perpendicular to area
density	1	-1	-1	-1	0	—
elastic modulus	1	1	-1	-1	-2	applied force
energy	1	2	0	0	-2	motion
force	1	1	0	0	-2	motion
frequency	0	0	0	0	-1	—
impulse	1	1	0	0	-1	applied force
length	0	1	0	0	0	length
mass	1	0	0	0	0	—
moment of inertia	1	0	1	1	0	perpendicular to axis
momentum	1	1	0	0	-1	motion
power	1	2	0	0	-3	applied force
pressure	1	1	-1	-1	-2	applied force
reciprocal length	0	-1	0	0	0	length
surface tension	1	1	$-\frac{1}{2}$	$-\frac{1}{2}$	-2	force
time	0	0	0	0	1	—
torque	1	0	1	1	-2	perpendicular to plane of torque
velocity	0	1	0	0	-1	motion
viscosity (dynamic)*	1	-1	0	0	-1	motion
viscosity, kinematic	0	0	1	1	-1	motion
volume	0	1	1	1	0	—

* To illustrate the construction of a vector dimensional form, consider viscosity η: it is defined by

$$\eta = \frac{F}{A} \bigg/ \frac{dv}{dz}.$$

F is the force applied tangential to an area A across the flow; v is the velocity of the flow and z the distance across the flow. Let the flow be in the [X]-direction.

F is applied in the [X]-direction: $\qquad\qquad\qquad\qquad\qquad$ $[F] = [M\ X\ T^{-2}]$

A has one axis in the [X]-direction and the other perpendicular to it (symmetrical in [Y], [Z]): $\qquad\qquad\qquad$ $[A] = [X\ Y^{\frac{1}{2}}\ Z^{\frac{1}{2}}]$

v is in the [X]-direction: $\qquad\qquad\qquad\qquad\qquad\qquad$ $[v] = [X\ T^{-1}]$

z is perpendicular to the [X]-direction (symmetrical in [Y], [Z]): $\quad$ $[z] = [Y^{\frac{1}{2}}\ Z^{\frac{1}{2}}]$

$$[\eta] = \frac{[M\ X\ T^{-2}]}{[X\ Y^{\frac{1}{2}}\ Z^{\frac{1}{2}}]} \bigg/ \frac{[X\ T^{-1}]}{[Y^{\frac{1}{2}}\ Z^{\frac{1}{2}}]}$$

$$= [M\ X^{-1}\ T^{-1}]$$

Angle can also be represented by the ratio of two different vector lengths, [X Y⁻¹]; it is generally possible to resolve the angle concerned in directions parallel to the appropriate axes. For example, a force acting in the horizontal plane might be considered as having two components F_x and F_y, where

$$[F_x] = [M \ X \ T^{-2}],$$
$$[F_y] = [M \ Y \ T^{-2}].$$

Alternatively, the resultant force is determined with reference to, say, the [X]-direction and written as

$$[F \cos \theta] = [M \ X \ T^{-2}].$$

It has also been proposed that mass may be described as either quantity of matter or inertial mass, and these two concepts distinguished by different dimensions. Even if this proposal were regarded as satisfactory, the resulting separation is unlikely to be particularly useful.

9 The method of dimensions

The application of dimensional forms to an analysis of problems by the principle of similitude, that all the terms in an equation must be dimensionally identical, is best explained by reference to a series of examples. For convenience, all the examples involve mechanical quantities.

It must be appreciated that the following limitations affect the application of the method of dimensional analysis.

(a) The formula is assumed to be of the form

$$f = k_1 \ x^{\alpha_1} \ y^{\beta_1} \ z^{\gamma_1} + k_2 \ x^{\alpha_2} \ y^{\beta_2} \ z^{\gamma_2} + \ldots,$$

where the value of k, α, β, γ are numerical (dimensionless) constants. It is not possible to separate out the individual terms on the right-hand side of the equation, and each has the same dimensional form.

(b) The sizes of the constants of proportionality k cannot be determined except by a rigorous theoretical proof. An experiment will yield their values to a certain degree of approximation.

(c) The conditions under which a formula may or may not hold cannot be included in the consideration. It is particularly evident that a variation of temperature affects the sizes of many quantities.

(d) No relevant quantity may be omitted from a dimensional analysis, although the ability to recognise all the quantities that are relevant can only be gained by experience.

Example A: A straightforward example with no more than the maximum number of solvable parameters (which is one more than the number of fundamental dimensions that are included).

Assume that the resistance to motion R of air flowing with a velocity v normal to an area A depends on v, A and the density ρ of the air according to the formula

$$R = k\rho^\alpha A^\beta v^\gamma.$$

Now,
$$[R] = [M\ L\ T^{-2}]\ \text{(a force)}$$
$$[\rho] = [M\ L^{-3}]$$
$$[A] = [L^2]$$
$$[v] = [L\ T^{-1}].$$

Thus
$$[M\ L\ T^{-2}] = [M\ L^{-3}]^\alpha\ [L^2]^\beta\ [L\ T^{-1}]^\gamma.$$

We equate the indices of the quantities [M], [L] and [T], which are totally independent of each other. This means that the dimensional equation is regarded as an identity.

For [M] $\qquad\qquad 1 = \alpha$
For [L] $\qquad\qquad 1 = -3\alpha + 2\beta + \gamma$
For [T] $\qquad -2 = -\gamma$

Solving these three equations simultaneously:

$$\alpha = 1$$
$$\gamma = 2$$
$$1 = -3 + 2\beta + 2 \ldots \beta = 1.$$

Thus the original equation should read

$$R = k\rho A v^2.$$

The value of k depends on the shape of the area. It is known to be approximately 0·3 for a circular area.

Example B: An example containing a superfluous parameter.
Assume that the period T of a simple pendulum depends on the length l of the pendulum, the mass m of the bob, and the acceleration of free fall g according to the formula

$$T = kl^\alpha m^\beta g^\gamma.$$

Now,
$$[T] = [T]$$
$$[l] = [L]$$
$$[m] = [M]$$
$$[g] = [L\ T^{-2}].$$

Thus
$$[T] = [L]^\alpha\ [M]^\beta\ [L\ T^{-2}]^\gamma.$$

Equating indices:

For [M] $\qquad\qquad$ $0 = \beta$
For [L] $\qquad\qquad$ $0 = a + \gamma$
For [T] $\qquad\qquad$ $1 = -2\gamma$

Solving these equations simultaneously:

$$\beta = 0$$
$$\gamma = -\tfrac{1}{2}$$
$$0 = a - \tfrac{1}{2} \ldots a = \tfrac{1}{2}.$$

Thus the original equation should read

$$T = k \, l^{\frac{1}{2}} \, g^{-\frac{1}{2}}$$
$$= k \sqrt{\frac{l}{g}}.$$

The value of k is known from rigorous theory to be 2π. The particular condition imposed on the application of this formula is that the amplitude of the oscillations is very small. The analysis has shown that the mass of the bob does not affect the period of the pendulum.

Example C: An example with more than the maximum number of solvable parameters.

Assume that the period T of a compound pendulum depends on the mass m of the pendulum, its moment of intertia I about an axis distant from the centre of gravity by h, and the acceleration of free fall g according to the formula

$$T = k \, I^{\alpha} \, m^{\beta} \, g^{\gamma} \, h^{\delta}.$$

Now,
$$[T] = [T]$$
$$[I] = [M \, L^2]$$
$$[m] = [M]$$
$$[g] = [L \, T^{-2}]$$
$$[h] = [L].$$

Thus
$$[T] = [M \, L^2]^{\alpha} \, [M]^{\beta} \, [L \, T^{-2}]^{\gamma} \, [L]^{\delta}.$$

Equating indices:
For [M] $\qquad\qquad$ $0 = a + \beta$
For [L] $\qquad\qquad$ $0 = 2a + \gamma + \delta$
For [T] $\qquad\qquad$ $1 = -2\gamma$

Solving these equations simultaneously:

$$\gamma = -\tfrac{1}{2}$$
$$\beta = -a$$
$$0 = 2a - \tfrac{1}{2} + \delta \ldots \delta = \tfrac{1}{2} - 2a.$$

Thus the original equation should read

$$T = k\, I^\alpha\, m^{-\alpha}\, g^{-\frac{1}{2}}\, h^{\frac{1}{2}-2\alpha}$$

$$= k\sqrt{\frac{h}{g}}\left(\frac{I}{m\,h^2}\right)^\alpha.$$

Further consideration of some sort would be necessary to show that

$$a = \tfrac{1}{2},$$

and that the formula should read

$$T = k\sqrt{\frac{I}{m\,g\,h}}.$$

The value of k is known from rigorous theory to be 2π. If we consider the case where $\alpha = 0$ (ie the moment of inertia does not affect the period), then

$$T = k\sqrt{\frac{h}{g}},$$

and the formula reduces to that for the simple pendulum, example **B**.

Example D: An example with no more than the maximum number of solvable parameters combined partly by addition. This case is of even greater importance when there are more than the maximum number of parameters. Consider an attempt to find the form of an equation in which a quantity f of dimensions $[M\ L^2\ T^{-3}]$ depends on the quantities

x, of dimensions $[L\ T^{-1}]$
y, of dimensions $[M\ T^{-1}]$
z, of dimensions $[M\ L\ T^{-2}]$.

The true form of the equation is known from rigorous theory to be

$$f = x^2 y + xz.$$

Assume the relationship to be in accordance with the formula

$$f = kx^\alpha y^\beta z^\gamma.$$

Thus $[M\ L^2\ T^{-3}] = [L\ T^{-1}]^\alpha\, [M\ T^{-1}]^\beta\, [M\ L\ T^{-2}]^\gamma.$

Equating indices:
For $[M]$ $1 = \beta + \gamma$
For $[L]$ $2 = \alpha + \gamma$
For $[T]$ $-3 = -\alpha - \beta - 2\gamma.$

It will be noted that these equations are not independent, the third being simply the sum of the first two. They therefore cannot be solved completely, but two of the unknowns must be put in terms of the other, γ say. Then

$$a = 2 - \gamma$$
$$\beta = 1 - \gamma.$$

Thus the equation should read

$$f = kx^{2-\gamma}y^{1-\gamma}z$$
$$= kx^2y\left(\frac{z}{xy}\right)^{\gamma}.$$

If $\gamma = 0$ $f = k_1x^2y;$
if $\gamma = 1$ $f = k_2xz.$

This means that the equation could be

$$f = k_1x^2y + k_2xz,$$

which conforms to the original statement, with

$$k_1 = k_2 = 1.$$

Example E: An example containing some parameters of a like physical nature.

Assume that the volume flow rate V/t of a liquid flowing in a streamline fashion along a tube of radius r and length l under a pressure p depends on r, l, p and the viscosity η of the liquid according to the formula

$$\frac{V}{t} = kp^{\alpha}r^{\beta}\eta^{\gamma}l^{\delta}.$$

Now, $[V/t] = [L^3\ T^{-1}]$
$$[p] = [M\ L^{-1}\ T^{-2}]$$
$$[r] = [L]$$
$$[\eta] = [M\ L^{-1}\ T^{-1}]$$
$$[l] = [L].$$

Thus $[L^3\ T^{-1}] = [M\ L^{-1}\ T^{-2}]^{\alpha}\ [L]^{\beta}\ [M\ L^{-1}\ T^{-1}]^{\gamma}\ [L]^{\delta}.$

Equating indices:

For [M] $0 = a + \gamma$
For [L] $3 = -a + \beta - \gamma + \delta$
For [T] $-1 = -2a - \gamma.$

Solving these equations simultaneously:

$$0 - 1 = a + \gamma - 2a - \gamma \ \ldots \ a = 1$$
$$0 = 1 + \gamma \qquad\qquad \ldots \ \gamma = -1$$
$$3 = -1 + \beta + 1 + \delta \ \ldots \ \delta = 3 - \beta.$$

Thus the original equation should read

$$\frac{V}{t} = kpr^\beta \eta^{-1} l^{3-\beta}$$

$$= k\frac{pl^3}{\eta}\left(\frac{r}{l}\right)^\beta.$$

From rigorous theory for this equation (Poiseuille's equation) it is known that

$$\beta = 4; \; k = \pi/8.$$

Thus

$$\frac{V}{t} = \frac{\pi}{8}\frac{pr^4}{\eta l}.$$

Example F: An example using dimensions of vector lengths.

The equation of example E will be treated by this method. Thus we assume

$$\frac{V}{t} = kp^\alpha r^\beta \eta^\gamma l^\delta.$$

Now, if the length of the tube is in the [X]-direction,

$$[V/t] = [X\ Y\ Z\ T^{-1}]$$
$$[p] = [M\ X\ Y^{-1}\ Z^{-1}\ T^{-2}]$$
$$[r] = [Y^{\frac12}\ Z^{\frac12}]$$
$$[\eta] = [M\ X^{-1}\ T^{-1}]$$
$$[l] = [X].$$

Thus $\quad [X\ Y\ Z\ T^{-1}] = [M\ X\ Y^{-1}\ Z^{-1}\ T^{-2}]^\alpha\ [Y^{\frac12}\ Z^{\frac12}]^\beta\ [M\ X^{-1}\ T^{-1}]^\gamma\ [X]^\delta.$

Equating indices:

For [M] $\qquad\qquad\qquad 0 = \alpha + \gamma$
For [X] $\qquad\qquad\qquad 1 = \alpha - \gamma + \delta$
For [Y] or [Z] $\qquad\quad 1 = -\alpha + \frac12\beta$
For [T] $\qquad\qquad\qquad -1 = -2\alpha - \gamma$

Solving these equations simultaneously:

$$0 - 1 = \alpha + \gamma - 2\alpha - \gamma \;\ldots\; \alpha = 1$$
$$0 = 1 + \gamma \qquad\qquad \ldots\; \gamma = -1$$
$$1 = -1 + \frac12\beta \qquad\quad \ldots\; \beta = 4$$
$$1 = 1 + 1 + \delta \qquad\quad \ldots\; \delta = -1$$

Thus the original equation should read

$$\frac{V}{t} = kpr^4\eta^{-1}l^{-1}$$

$$= k\frac{pr^4}{\eta l},$$

which is Poiseuille's equation as stated at the end of example E.

It may be useful, finally, to summarise the advantages gained from a knowledge of dimensions and the methods of dimensional analysis.

(a) The general (but not necessarily complete) solution of a complex equation can be determined by simple theory. It may be impossible to obtain a complete solution by rigorous analysis. Alternatively, an experimenter can be guided to a full experimental solution by a knowledge of the functional relationships predicted by dimensional analysis.

(b) An equation derived by analysis can be checked at each stage to reveal possible errors or to confirm imperfectly-remembered relationships.

(c) The dimensional form of a quantity can be used to provide a conversion value from one set of units to another. Eg density ρ in grammes per centimetre cubed (g/cm³) can be obtained from the value in kilogrammes per metre cubed (kg/m³) using the knowledge that

$$1 \text{ cm} = 10^{-2} \text{ m}$$
$$1 \text{ g} = 10^{-3} \text{ kg}$$
and
$$[\rho] = [\text{M L}^{-3}].$$
The conversion factor needed $= 10^{-3} (10^{-2})^{-3}$
$$= 10^{3}.$$

The value of the density in kg/m³ must be divided by 10^3 to obtain the value in g/cm³.

(d) The principle of similitude has been used as an aid to making scale models for experimental analysis in wind tunnels, etc.

10 Tabular lists

The following tables list the dimensions of all the quantities to be found in part 2 (Quantities) arranged in ascending order of the values of the indices.

Table 19: The dimensions of mechanical and acoustical quantities, and those of thermal and photometric quantities in the MLT system

Note: (H) = MLT dimensions as applied to a quantity in heat
(L) = MLT dimensions as applied to a quantity in light

M	L	T	
−1	0	2	compliance (mechanical)
−1	1	1	fluidity; (H) thermal resistivity
−1	1	2	compressibility
−1	2	−1	(H) thermal resistance (2)
−1	2	0	mass absorption coefficient, mass attenuation coefficient, specific area; (H) thermal inductance
−1	2	1	(H) thermal resistance (1)
−1	3	0	specific volume

M	L	T	
−1	4	2	acoustical compliance
0	−2	2	(H) temperature coefficient (eg expansion, pressure)
0	−1	0	circular wave number, curvature, linear absorption coefficient, linear attenuation coefficient, wave number
0	0	−2	angular acceleration
0	0	−1	activity, angular velocity, damping coefficient, frequency, rotational frequency, velocity potential
0	0	0	(see table 20)
0	0	1	time
0	1	−2	acceleration, gravitational field strength; (H) temperature gradient
0	1	−1	velocity
0	1	0	aperture conductivity, length, etc
0	2	−2	gravitational potential, specific energy; (H) temperature, specific latent heat
0	2	−1	diffusion coefficient, kinematic viscosity; (H) thermal diffusivity
0	2	0	area, fuel consumption traffic factor
0	3	−1	sound strength, volume flow rate
0	3	0	section modulus, volume
0	4	0	second moment
1	−4	−2	acoustical stiffness
1	−4	−1	acoustical impedance, acoustical reactance, acoustical resistance
1	−4	0	acoustical mass
1	−3	0	absolute humidity, concentration, density; (H) specific heat capacity (volume basis)
1	−2	−2	specific weight
1	−2	−1	specific acoustical impedance, specific acoustical reactance, specific acoustical resistance; (H) thermal conductance
1	−2	0	area density, mass per fuel consumption traffic factor
1	−2	2	(H) thermal capacitance
1	−1	−3	(H) heat release
1	−1	−2	elastic modulus, energy density, pressure, resilience, stress; (H) calorific value (volume basis)
1	−1	−1	dynamic viscosity; (H) thermal conductivity, thermal current
1	−1	0	line density
1	0	−3	sound intensity; (electricity:) Poynting vector; (H) heat flow rate density, irradiance, radiance, radiant emittance; (L) illumination, luminance, luminosity, luminous emittance, point brilliance, equivalent luminance
1	0	−2	stiffness (mechanical), surface tension
1	0	−1	mass flow rate, mechanical impedance, mechanical reactance, mechanical resistance
1	0	0	mass; (H) entropy, heat capacity
1	1	−2	force, weight; (H) spectral radiant energy
1	1	−1	impulse, translational momentum
1	1	0	mass-distance traffic factor
1	2	−3	power, sound flux; (H) heat flow rate, radiant flux, radiant intensity; (L) luminous flux, luminous intensity
1	2	−2	energy, etc, moment, torque
1	2	−1	action, angular momentum
1	2	0	moment of inertia

Table 20: Dimensionless [M⁰ L⁰ T⁰] physical quantities

absorption coefficient	Q-factor
absorptivity	quality factor
amplitude level	reflectivity
attenuation coefficient	reflexion coefficient
coupling coefficient (induction)	refractive index
density of a transmitting medium	relative density
dissipation coefficient	relative humidity
efficiency	relative luminous efficiency
electric susceptibility	relative permeability
equivalent loudness	relative permittivity
extinction coefficient	restitution coefficient
friction coefficient	saturation ratio
hardness	scattering coefficient
intensity level	sensation level
interval	solid angle
leakage coefficient (induction)	sound pressure level
logarithmic decrement	specific gravity
loudness	specific humidity
luminance factor	specific viscosity
magnetic susceptibility	strain
mechanical advantage	transmission coefficient
mixing ratio	transmissivity
opacity	transmittance
perceived noise level	transmittancy
plane angle	vapour density
Poisson's ratio	velocity ratio
principal specific heat capacities ratio	

Table 21: The dimensions of thermal quantities in the MLTQΘ system

M	L	T	Q	Θ	
−1	0	0	1	−1	specific entropy, specific heat capacity (mass basis)
−1	0	0	1	0	specific latent heat
0	−3	−1	1	0	heat release
0	−3	0	1	−1	specific heat capacity (volume basis)
0	−3	0	1	0	calorific value (volume basis)
0	−2	−1	1	−1	thermal conductance
0	−2	−1	1	0	heat flow rate density, irradiance, radiance, radiant emittance
0	−1	−1	1	−1	thermal conductivity, thermal current
0	−1	0	0	1	temperature gradient
0	−1	0	1	0	spectral radiant energy
0	0	−1	1	0	radiant flux, radiant intensity
0	0	0	0	−1	temperature coefficient (eg expansion, pressure)
0	0	0	0	1	temperature
0	0	0	1	−2	thermal capacitance
0	0	0	1	−1	entropy, heat capacity
0	0	0	1	0	heat energy
0	0	1	−1	2	thermal resistance (2)
0	0	2	−1	2	thermal inductance
0	1	1	−1	1	thermal resistivity
0	2	1	−1	1	thermal resistance (1)

Table 22: The dimensions of photometric quantities in the MLTF system

M	L	T	F	
−1	−2	3	1	luminous efficiency
0	−2	0	1	equivalent luminance, illumination, luminance, luminosity, luminous emittance, point brilliance
0	0	1	1	luminous energy
0	0	0	1	luminous flux, luminous intensity

Table 23: The dimensions of electrical and magnetic quantities in the MLTI system

M	L	T	I	
−1	−3	3	2	conductivity
−1	−3	4	2	absolute permittivity
−1	−2	2	2	reluctance
−1	−2	3	2	admittance, conductance, susceptance
−1	−2	4	2	capacitance
−1	0	1	1	specific charge
0	−3	1	1	volume density of charge
0	−2	0	1	current density
0	−2	1	1	electric polarisation, electric displacement, surface density of charge
0	−1	0	1	linear current density, magnetic field strength, magnetisation
0	0	0	1	current, magnetic potential, magnetomotive force
0	0	1	1	charge, electric flux
0	2	0	1	electromagnetic moment
1	0	−3	0	Poynting vector
1	0	−2	−1	magnetic flux density, magnetic polarisation
1	1	−3	−1	electric field strength, electrisation
1	1	−2	−2	absolute permeability
1	1	−2	−1	magnetic vector potential
1	1	1	1	electric dipole moment
1	2	−4	−2	elastance
1	2	−3	−2	impedance, reactance, resistance
1	2	−3	−1	electric potential, electromotive force, Peltier coefficient, Seebeck coefficient, Thomson coefficient,* thermoelectric power,* voltage
1	2	−2	−2	inductance, permeance
1	2	−2	−1	magnetic flux, magnetic pole strength
1	3	−3	−2	resistivity
1	3	−2	−1	magnetic dipole moment
2	0	−3	−2	mass resistivity

* Additionally, the dimensional form requires Θ^{-1}

Table 24: The dimensions of electrical and magnetic quantities in the MLTμ system

M	L	T	μ	
−$\frac{1}{2}$	$\frac{1}{2}$	0	−$\frac{1}{2}$	specific charge
0	−2	1	−1	conductivity
0	−2	2	−1	absolute permittivity

M	L	T	μ	
0	-1	0	-1	reluctance
0	-1	1	-1	admittance, conductance, susceptance
0	-1	2	-1	capacitance
0	0	0	1	absolute permeability
0	1	-2	1	elastance
0	1	-1	1	impedance, reactance, resistance
0	1	0	1	inductance, permeance
0	2	-1	1	resistivity
$\frac{1}{2}$	$-\frac{5}{2}$	0	$-\frac{1}{2}$	volume density of charge
$\frac{1}{2}$	$-\frac{3}{2}$	-1	$-\frac{1}{2}$	current density
$\frac{1}{2}$	$-\frac{3}{2}$	0	$-\frac{1}{2}$	electric displacement, electric polarisation, surface density of charge
$\frac{1}{2}$	$-\frac{1}{2}$	-1	$-\frac{1}{2}$	linear current density, magnetic field strength, magnetisation
$\frac{1}{2}$	$-\frac{1}{2}$	-1	$\frac{1}{2}$	magnetic flux density, magnetic polarisation
$\frac{1}{2}$	$\frac{1}{2}$	-2	$\frac{1}{2}$	electric field strength, electrisation
$\frac{1}{2}$	$\frac{1}{2}$	-1	$-\frac{1}{2}$	current, magnetic potential, magnetomotive force
$\frac{1}{2}$	$\frac{1}{2}$	-1	$\frac{1}{2}$	magnetic vector potential
$\frac{1}{2}$	$\frac{1}{2}$	0	$-\frac{1}{2}$	charge, electric flux
$\frac{1}{2}$	$\frac{3}{2}$	-2	$\frac{1}{2}$	electric potential, electromotive force, Peltier coefficient, Seebeck coefficient, Thomson coefficient,* thermoelectric power,* voltage
$\frac{1}{2}$	$\frac{3}{2}$	-1	$\frac{1}{2}$	magnetic flux, magnetic pole strength
$\frac{1}{2}$	$\frac{3}{2}$	0	$-\frac{1}{2}$	electric dipole moment
$\frac{1}{2}$	$\frac{5}{2}$	-1	$-\frac{1}{2}$	electromagnetic moment
$\frac{1}{2}$	$\frac{5}{2}$	-1	$\frac{1}{2}$	magnetic dipole moment
1	-1	-1	1	mass resistivity
1	0	-3	0	Poynting vector

* Additionally, the dimensional form requires Θ^{-1}

Table 25: The dimensions of electrical and magnetic quantities in the MLTε system

M	L	T	$\in$	
$-\frac{1}{2}$	$\frac{3}{2}$	-1	$\frac{1}{2}$	specific charge
0	-2	2	-1	absolute permeability
0	-1	0	-1	elastance
0	-1	1	-1	impedance, reactance, resistance
0	-1	2	-1	inductance, permeance
0	0	-1	1	conductivity
0	0	0	1	absolute permittivity
0	0	1	-1	resistivity
0	1	-2	1	reluctance
0	1	-1	1	admittance, conductance, susceptance
0	1	0	1	capacitance
$\frac{1}{2}$	$-\frac{3}{2}$	-1	$\frac{1}{2}$	volume density of charge
$\frac{1}{2}$	$-\frac{3}{2}$	0	$-\frac{1}{2}$	magnetic flux density, magnetic polarisation
$\frac{1}{2}$	$-\frac{1}{2}$	-2	$\frac{1}{2}$	current density

M	L	T	$\in$	
$\frac{1}{2}$	$-\frac{1}{2}$	-1	$-\frac{1}{2}$	electric field strength, electrisation
$\frac{1}{2}$	$-\frac{1}{2}$	-1	$\frac{1}{2}$	electric displacement, electric polarisation, surface density of charge
$\frac{1}{2}$	$-\frac{1}{2}$	0	$-\frac{1}{2}$	magnetic vector potential
$\frac{1}{2}$	$\frac{1}{2}$	-2	$\frac{1}{2}$	linear current density, magnetic field strength, magnetisation
$\frac{1}{2}$	$\frac{1}{2}$	-1	$-\frac{1}{2}$	electric potential, electromotive force, Peltier coefficient, Seebeck coefficient, Thomson coefficient,* thermoelectric power,* voltage
$\frac{1}{2}$	$\frac{1}{2}$	0	$-\frac{1}{2}$	magnetic flux, magnetic pole strength
$\frac{1}{2}$	$\frac{3}{2}$	-2	$\frac{1}{2}$	current, magnetic potential, magnetomotive force
$\frac{1}{2}$	$\frac{3}{2}$	-1	$\frac{1}{2}$	charge, electric flux
$\frac{1}{2}$	$\frac{3}{2}$	0	$-\frac{1}{2}$	magnetic dipole moment
$\frac{1}{2}$	$\frac{5}{2}$	-1	$\frac{1}{2}$	electric dipole moment
$\frac{1}{2}$	$\frac{7}{2}$	-2	$\frac{1}{2}$	electromagnetic moment
1	-3	1	-1	mass resistivity
1	0	-3	0	Poynting vector

* Additionally, the dimensional form requires Θ^{-1}

Appendix 3

METRIC MULTIPLES AND SUBMULTIPLES

1 Powers of ten

Table 26 gives the names used for powers of ten in the United Kingdom (UK) and the United States (US). The UK names are also used in the countries of Europe and, with the exception of the milliard, these are the names recommended at the 9th CGPM (Conférence Générale des Poids et Mesures) of 1948 for international use. The naming of powers of ten higher than those tabulated follows the same system.

Table 26: Powers of ten

size		UK name	US name
10 $=$	10	ten	ten
10^2 $=$	100	hundred	hundred
10^3 $=$	$1\ 000$	thousand	thousand
10^6 $=$	$1\ 000\ 000$	million	million
10^9 $=$	$1\ 000\ 000\ 000$	milliard	billion
10^{12} $=$	$1\ 000\ 000\ 000\ 000$	billion	trillion
10^{15} $=$	$1\ 000\ 000\ 000\ 000\ 000$	—	quadrillion
10^{18} $=$	$1\ 000\ 000\ 000\ 000\ 000\ 000$	trillion	quintillion

2 Prefixes denoting multiples and submultiples

Table 27 gives the prefixes used for multiples and submultiples of ten. With the exception of beva-, myria-, femto- and atto-, all the prefixes were accepted for international use by the 11th CGPM of 1960; the last two were accepted for international use by the 12th CGPM of 1964. Beva- (which is used only in the US with certain units, eg the electronvolt) and myria- (which has been used in France with certain units, eg the metre) are not recommended. The prefix symbol is written next to the unit symbol with no gap between them.

Notes

(a) The shorthand notation for writing small values is rarely used, and can be confusing.

(b) It is recommended that, as far as possible, the units of the SI (international system of units) should be modified by prefixes that represent steps of 10^3 and 10^{-3} only. Thus the use of myria-, hecto-, deca-, deci- and centi- are not recommended.

Table 27: Prefixes for multiples and submultiples of ten

prefix	symbol	size	shorthand notation
tera-	T	$10^{12} = 1\ 000\ 000\ 000\ 000$	
giga-	G		
beva-	B	$10^9 = 1\ 000\ 000\ 000$	
mega-	M	$10^6 = 1\ 000\ 000$	
myria-	my	$10^4 = 10\ 000$	
kilo-	k	$10^3 = 1\ 000$	
hecto-	h	$10^2 = 100$	
deca-	da	$10 = 10$	
deci-	d	$10^{-1} = 0\cdot1$	
centi-	c	$10^{-2} = 0\cdot01$	
milli-	m	$10^{-3} = 0\cdot001$	$0\cdot0^2\ 1$
micro-	μ	$10^{-6} = 0\cdot000\ 001$	$0\cdot0^5\ 1$
nano-	n	$10^{-9} = 0\cdot000\ 000\ 001$	$0\cdot0^8\ 1$
pico-	p	$10^{-12} = 0\cdot000\ 000\ 000\ 001$	$0\cdot0^{11}\ 1$
femto-	f	$10^{-15} = 0\cdot000\ 000\ 000\ 000\ 001$	$0\cdot0^{14}\ 1$
atto-	a	$10^{-18} = 0\cdot000\ 000\ 000\ 000\ 000\ 001$	$0\cdot0^{17}\ 1$

(c) When writing the unit and prefix in full, the final vowel of the prefix is generally retained even if the unit itself begins with a vowel, although a hyphen is employed when the vowels are the same: thus deca-ampère, milli-inch, but millioersted. The exceptions to this rule are:

microhm (not micro-ohm)   decare (not deca-are)
kilohm (not kilo-ohm)   hectare (not hectoare)
megohm (not megaohm)   kiliare (not kiloare)

(d) In the case of a compound unit, the prefix represents a change in size of the whole unit and not just that part to which it is attached, ie the attachment of a prefix effectively constitutes a new unit. Thus

$$1\ \mu m^2 = 1\ (\mu m)^2 = 10^{-12}\ m^2;\ \text{and not}$$
$$1\ \mu m^2 = 1\ \mu(m^2) = 10^{-6}\ m^2.$$

(e) The use of a double prefix is deprecated. Eg ng should be used in place of mμg. In the past the prefix dimilli- (dm) has been used for the submultiple $10^{-4} = 0\cdot000\ 1$. An example of this is dmsb (dimillistilb), which is better written $0\cdot1$ msb ($0\cdot1$ millistilb).

(f) The gramme is likely to remain the named basic unit of mass, even though the kilogramme is the fundamental unit of mass in the SI. Despite this, multiples and submultiples of the kilogramme should be written as multiples and submultiples of the gramme: thus mg and not μkg.

(g) The prefix representing ten times was formerly deka- (dk).

3 Recommended multiples and submultiples of units

Table 28 gives those multiples and submultiples of units, both SI and others, that are most likely to be required. By using the recommendations given, the multiplicity of possibilities is cut down.

The main body of the table indicates the multiples and submultiples to be given to the unit, or to the numerator of a compound unit. Under the heading 'other multiples' will be found examples where it is common practice to give the multiples and submultiples to the denominator of a compound unit (a course that is strictly not regarded as good), together with a few unusual combinations, eg combinations that include a unit of time other than the second.

Table 28: Recommended multiples and submultiples of units

unit	multiple or submultiple													other multiples
	T	G	M	k	h	da	-1	d	c	m	μ	n	p	
SI UNITS														
ampère				*		°	*			*	*	*	*	
ampère per metre				*			*							ampère per millimetre; ampère per centimetre
ampère per metre squared			+	*			*			*	*	*		ampère per millimetre squared; ampère per centimetre squared
ampère per metre cubed			+				*			*	*			ampère per millimetre cubed; ampère per centimetre cubed
candela			+				*							
coulomb				+			*			*	*	*	*	
coulomb metre				*			*							coulomb centimetre
coulomb per metre squared			+	*			*			+	*	*	*	coulomb per millimetre squared; coulomb per centimetre squared
coulomb per metre cubed			+	*			*							coulomb per millimetre cubed; coulomb per centimetre cubed
farad							+			*	*	+	*	
farad per metre							*			*	*	+	+	
henry							*			*	*	+		
henry per metre							*				*			
hertz	+	+	*	*			*							
joule	+	+	*	*			*			*				
joule per kelvin				*			*							
joule per kilogramme			*	*			*							
joule per kilogramme kelvin				*			*							
joule per metre squared				*			*							joule per centimetre squared; decajoule per centimetre squared
kilogramme²							*							

Unit	Related units
kilogramme per metre cubed	kilogramme per decimetre cubed[3]
metre	
metre squared	
metre cubed	
metre to the power of four	
metre per second	kilometre per hour
metre squared per second	
metre cubed per second	metre cubed per minute; metre cubed per hour
newton	
newton metre	
newton second per metre squared	
newton per metre	
newton per metre squared[5]	newton per millimetre squared; newton per centimetre squared; kilonewton per millimetre squared; decanewton per millimetre squared
ohm	
ohm metre	ohm centimetre; teraohm centimetre; milliohm centimetre; microhm centimetre
pascal	
second	day; hour; minute
siemens	
siemens per metre	
tesla	
var	
volt	
volt per metre	volt per millimetre; volt per centimetre
voltampère	
watt	

unit	T	G	M	k	h	da	$^{-1}$	d	c	m	μ	n	p	other multiples
watt second[6]	+	+	*	*			*			*				watt hour; kilowatt hour; megawatt hour; gigawatt hour; terawatt hour
watt per metre squared			*	*			*			*				
weber							*							
weber per metre				+			*							weber per millimetre
CGS UNITS														
dyne			*				*			*				
galileo (Gal)			*	*			*	o	o	*	*			
gramme							*							
gramme per centimetre cubed							*7			*				megagramme per metre cubed
phot							*	o		*				
poise							*		o	*				
stokes							*		o	*				
OTHER METRIC UNITS														
are				+	o		*							
bar				+	o		o							
barn							*							
gramme per litre				*			*			*	*			gramme per millilitre[8]
gramme per mole				*			*			*				
gramme-force			+	*			*							
joule per mole					o		*							joule per kilomole
joule per mole kelvin					o		*							joule per kilomole kelvin
lambert					o		*			*				
litre				o			*		o	*				
litre per second							*							litre per hour; litre per minute

metre cubed per kilomole

mole per decimetre cubed

kilocalorie per kilogramme
calorie per centimetre cubed
kilocalorie per hour
kilocalorie per kilogramme kelvin
kilocalorie per metre squared hour;
   calorie per centimetre squared second
kilocalorie per metre hour kelvin; calorie
   per centimetre second kelvin
kilocalorie per metre squared hour
   kelvin; calorie per centimetre squared
   second kelvin

---

lusec
metre cubed per mole
micron
mole
mole per kilogramme
mole per metre cubed
mole per litre
pièze

OTHER UNITS
bel
calorie
calorie per gramme
calorie per metre cubed
calorie per second
calorie per gramme kelvin
calorie per metre squared second

calorie per metre second kelvin

calorie per metre squared second
   kelvin

calorie per metre cubed
calorie per metre cubed kelvin
curie
electronvolt
gramme-röntgen
inch
k
nile

unit	multiple or submultiple													other multiples
	T	G	M	k	h	da	1	d	c	m	μ	n	p	
parsec			*	*			*								
radian							*								
rayleigh		*		*			*								
revolution per second							*				*	+			revolution per minute
röntgen							*				*	*			
Siegbahn unit			*				*								
ton (explosive power)				*			°								
torr				*			*				*				

* Especially recommended
+ Recommended
° In general use
1 Use of the unit without a prefix
2 See gramme
3 = gramme per centimetre cubed
  = gramme per millilitre
4 = litre
5 = pascal
6 = joule
7 = kilogramme per decimetre cubed
  = gramme per millilitre
8 = kilogramme per decimetre cubed
  = gramme per centimetre cubed

Appendix 4

OBSOLETE AND OLD-FASHIONED UNITS

Table 29 lists a large number of imperial units (and a few metric units), most of which are employed commercially for the measurement of particular commodities in the United Kingdom (UK) and the United States (US), and almost all of which are going out of use or are already obsolete.

In the first column ('unit') a dagger indicates that an entry will be found in part 1 (Units).

In the second column ('type') the following abbreviations are employed:

A	area	M	mass
C	capacity	V	volume
L	length		

In the third column ('size') an asterisk indicates that there exist local variations in the size of the unit. The unit symbols employed are:

bu	bushel	lb	pound
cwt	hundredweight	lb tr	troy pound
ft	foot	oz	ounce
gal	gallon	pt	pint
gr	grain	WWG	'Winchester' wine gallon
in	inch	yd	yard
l	litre		

Table 29: Obsolete and old-fashioned units

unit	type	size	commodity, etc; (notes)
acre-foot	V	43 560 ft^3	irrigation engineering
acre-inch	V	3 630 ft^3	irrigation engineering
anker	C	10 WWG	honey, oil, spirits, vinegar, wine
aum	C	30 gal	hock
bag	C	4 or 5 bu	potatoes
	M	28 to 316 lb	pepper
		49 or 98 lb	flour (US)
		50 lb	vermillion
		94 lb	cement (US)
		*1 cwt	cocoa (UK)
		1 or 1$\frac{1}{2}$ cwt	potatoes
		1 to 1$\frac{1}{4}$ cwt	ginger
		1 to 1$\frac{1}{2}$ cwt	Mauritias sugar
		1 to 1$\frac{3}{4}$ cwt	East Indies sugar
		1$\frac{1}{4}$ to 1$\frac{1}{2}$ cwt	coffee (UK)
		*1$\frac{1}{2}$ cwt	rice
		1$\frac{1}{2}$ to 1$\frac{3}{4}$ cwt	Messina nuts
		200 lb	cochineal (UK)
		2$\frac{1}{2}$ cwt	flour, hops (UK)
		364 lb	wool (UK)
bale	M	92$\frac{1}{2}$ lb	cinnamon
		180 to 500 lb	cotton wool
		2 to 2$\frac{1}{2}$ cwt	Mocha coffee (UK)
barge	M	21$\frac{1}{6}$ tons	coal; (= keel)
barleycorn	L	$\frac{1}{3}$ in	
barn gallon: see gallon, barn			
barony	A	4 000 acres	land
barrel	C	26$\frac{1}{2}$ gal	tar
		32 gal	herrings
		36 WWG	ale, beer
		31$\frac{1}{2}$ USgal	liquids
		42 USgal	petrol (9 702 in^3)
	M	30 lb	anchovies (UK)
		96 to 360 lb	figs (UK)
		1 cwt	gunpowder, raisins (UK)
		1 to 1$\frac{1}{2}$ cwt	coffee (UK)
		120 lb	candles, potash, potatoes (UK)
		*1$\frac{1}{4}$ cwt	tapioca (UK)
		180 lb	lime, small (US)
		196 lb	flour (US)
		200 lb	beef, fish, pork (US)
		2 cwt	butter
		*2 cwt	resin (UK)
		2 to 2$\frac{1}{2}$ cwt	turpentine (UK)

unit	type	size	commodity, etc; (notes)
barrel		256 lb	soap (UK)
		280 lb	lime, large (US)
		376 lb	cement (US)
	V	5 826 in³	cranberries (US)
		7 056 in³	fruit, vegetables (US)
		9 702 in³	petrol (42 USgal) (US)
barrel bulk	V	5 ft³	freight
barrique	C	225 l	claret wine
basket	M	1¼ to 1½ cwt	almonds
blank	M	1/230 400 gr	moneyweight
board foot (bd ft)	V	144 in³	timber
bodge	C	½ gal	(= pottle, quarter, quartern)
boll	M	140 lb	flour
bolt	L	16 yd	wallpaper (US)
		40 yd	cloth (US)
bottle	M	*86 lb	mercury
box	M	22 lb	Malaga raisins
		25 lb	Jourdan almonds
		30 to 40 lb	Valencia raisins
		*1 cwt	camphor
†bushel	M	*40 lb	apples
bushel, international corn	M	60 lb	wheat
butt	C	108 gal	ale, beer
		252 WWG	wine
	M	15 to 20 cwt	currants
carton	M	9 lb	plums
case	M	*1½ cwt	mace
cask	M	9 to 18 lb	mustard
		1 cwt	Malaga raisins
		1¼ cwt	cocoa
		200 lb	nutmegs
		2½ cwt	Turkey raisins
		3 to 4 cwt	soda
		6 cwt	rice (US)
		*9 cwt	tallow
†chaldron	C	36 bu	coke, etc
	M	25½ cwt	coal
chest	C	125 gal	olive oil
	M	*110 lb	tea
		136 lb	Turkey opium
		149⅓ lb	East Indies opium
		200 lb	cloves
		4 to 6 cwt	gum Arabic
clove	M	7 lb	wool
		8 lb	butter, cheese
comb = coom(b)	C	4 bu	
cord	V	128 ft³	timber

unit	type	size	commodity, etc; (notes)
Corn bushel, international: see bushel, international corn			
cran(e)	C	37½ gal	fresh herrings
curnock	C	3 or 4 bu	
cut	L	300 yd	linen yarn
digit	L	¾ in	
double sack: see sack, double			
douzième	L	1/144 in	
droit	M	1/480 gr	moneyweight
drop	M	$\frac{1}{16}$ oz	spirits (Scottish)
drum	C	50 to 55 USgal	petrol (US)
	M	*24 lb	raisins
ell, English	L	45 in	cloth
ell, Flemish	L	27 in	cloth
ell, French	L	54 in	cloth
fag(g)ot	M	120 lb	steel
fangot	M	1 to 3 cwt	raw silks, etc
fathom (cubic)	V	216 ft³	mining engineering
fathom (square)	A	36 ft²	mining engineering
firkin	C	9 gal	ale, beer (UK)
		9·8 USgal	ale, beer (US)
	M	56 lb	butter
		64 lb	soap
flask	M	75 lb	mercury (US)
fodder	M	2 400 lb	lead; (= tother)
fother	M	2 184 lb	lead, etc (varies between 19 and 24 cwt)
fotmal	M	*70 lb	lead, etc
†gallon	M	9 lb	train oil
		12 lb	honey
	V	282 in³	ale, beer
gallon, barn	C	2 gal	milk (UK)
		2·4 USgal	milk (US)
gallon, 'Winchester' wine	C	0·833 1 UKgal	honey, oil, spirits, vinegar, wine
goad	L	4½ ft	
great pound: see pound, great			
hand	L	4 in	heights of horses
handbreadth	L	2½ to 4 in	
hank	L	840 yd	cotton yarns
heer	L	600 yd	linen and woollen yarns
hide	A	100 acres	land
hogshead (hhd)	C	40 gal	pilchards
		45 to 50 gal	rum
		45 to 60 gal	brandy
		54 gal	ale, beer
		55 to 60 gal	whisky
		63 WWG	honey, oil, spirits, vinegar, wine

unit	type	size	commodity, etc; (notes)
hogshead (hhd)	M	12 to 18 cwt	tobacco
		13 to 16 cwt	sugar
international corn bushel: see bushel, international corn			
iron	L	1/48 in	leather (US)
†jar	C	25 gal	olive oil
keel	M	$21\frac{1}{5}$ tons	coal; (= barge)
keg	C	4 to 5 gal	sturgeons
	M	100 lb	nails (US)
kilderkin	C	18 gal	ale, beer
last	C	80 bu	grain, etc
	M	1 700 lb	feathers, flax
		2 400 lb	gunpowder
		39 cwt	wool
lea	L	120 yd	cotton yarn
†line	L	1/12 in	(= second)
		1/40 in	button diameters (US)
load	A	600 ft²	1-inch planks; (= ton)
	C	40 bu	(= wey)
	M	11 cwt 64 lb	straw
		18 cwt	old hay
		19 cwt 32 lb	new hay
	V	40 ft³	unhewn timber; (= ton)
		50 ft³	timber; (= ton)
mast	M	2·5 lb tr	
matt	M	80 lb	cloves
		1 to 1½ cwt	sugar; (= bag)
measurement ton: see ton			
military pace: see pace, military			
mite	M	1/20 gr	moneyweight
nail	L	$2\frac{1}{4}$ in	cloth
	M	7 lb	wool; (= clove)
		8 lb	butter, cheese; (= clove)
noggin	C	$\frac{1}{4}$ pt	liquids
†ounce	L	1/64 in	leather
pace	L	5 ft	
pace, military	L	$2\frac{1}{2}$ ft	
pack	M	240 lb	wool
palm	L	3 in	
†peck	M	14 lb	flour
perit	M	1/9 600 gr	moneyweight
piece	L	10 yd	muslin
		28 yd	calico
pin	C	$4\frac{1}{2}$ gal	ale, beer
pipe	C	100 to 118 gal	cider
		126 WWG	ale, beer; (= butt)
pocket	M	168 to 224 lb	hops
		182 lb	wool

unit	type	size	commodity, etc; (notes)
point	L	1/76 in	type
		0·001 in	paper thickness
pottle	C	4 pt	(= bodge, quarter, quartern)
pound, great	M	24 oz	some silks
puncheon	C	72 gal	ale, beer
		90 to 100 gal	rum
		100 to 110 gal	brandy
		112 to 120 gal	whisky
		84 WWG	honey, oil, spirits, vinegar, wine
	M	10 to 12 cwt	molasses
quart, reputed	C	$\frac{1}{6}$ WWG	wine
†quarter	C	4 pt (= ¼ peck)	(= bodge, pottle, quartern)
		8 bu	grain, etc; (= seam)
	L	9 in (= ¼ yd)	cloth
	M	3½ lb (= ¼ stone)	
quartern	C	4 pt	(= bodge, pottle, quarter)
register ton: see ton			
reputed quart: see quart, reputed			
room	M	7 tons	coal
run(d)let	C	18 WWG	honey, oil, spirits, vinegar, wine
sack	C	3 bu	coke (UK)
		5 bu	flour, salt (UK)
	M	28 lb	cotton (UK)
		100 lb	flour, meal (US)
		112 lb	coal (UK)
		140 lb	cotton, flour for export (US)
		215 lb	salt (US)
		280 lb	flour, meal (UK)
		364 lb	wool (UK)
sack, double	M	2 cwt	coal
score	M	20 to 21 lb	wool
seam	C	8 bu	grain; (= quarter)
	M	120 lb	glass
†second	L	1/12 in	(= line)
seron	M	140 lb	cochineal
		1¼ to 2 cwt	almonds
		250 lb	indigo
shipload	M	424 tons	coal
shipping ton: see ton			
skein	L	360 ft	cotton yarn
span	L	9 in	
spindle	L	14 400 yd	linen yarn
		15 120 yd	cotton yarn
square (of flooring)	A	100 ft²	building, etc
stack	V	108 ft³	fuel, timber

unit	type	size	commodity, etc; (notes)
stand	M	2½ to 3 cwt	pitch
standard	V	*16⅔ ft³	timber (US)
		165 ft³	timber
†stone	M	5 lb	glass
		8 to 20 lb	meat (standard stone: 8 lb)
		12 to 16 lb	wool (standard stone: 14 lb)
		16 lb	cheese
		32 lb	hemp
strike	C	½ to 4 bu	
thread	L	1½ yd	cotton yarn
tierce	C	42 WWG	honey, oil, spirits, vinegar, wine
	M	304 lb	Irish beef
		320 lb	Irish pork
		5 to 7 cwt	coffee
		7 to 9 cwt	sugar
tod	M	28 lb	wool
†ton	A	60 ft²	1-inch planks; (= load)
	C	1 770 gal	oil
	V	40 ft³	sea freight (also called measurement ton, shipping ton); unhewn timber (= load)
		50 ft³	timber; (= load)
		100 ft³	internal capacity of ships (also called register ton)
tother	M	2 400 lb	lead; (= fodder)
trug	C	⅔ bu	wheat
truss	M	36 lb	straw
		56 lb	old hay
		60 lb	new hay
tub	M	84 lb	butter
tun	C	210 gal	oil
		252 WWG	honey, oil, spirits, vinegar, wine
watch	time	4 hours	ship's time
wey	C	40 bu	(= load)
		182 lb	wool
		256 or 336 lb	butter, cheese
winchester quart	C	2½ l	liquids
'Winchester' wine gallon: see gallon, 'Winchester' wine			
yard, cubic	M	6 stones	new hay
		8 stones	old hay
yard of land	A	30 acres	land

Appendix 5

TABLES OF THE RELATIONS BETWEEN UNITS THAT MEASURE THE SAME QUANTITIES

1 Introduction

In general the following tables of unit relationships exclude any reference to units the sizes of which depend on experimentally-determined quantities and units that are of an indefinite size. The obsolete and old-fashioned units given in Appendix 4 for the measurement of particular commodities are also excluded. Some of the units have alternative names, and these are quoted under the entries in part 1 (Units).

2 Angle

Table 30: Units of angle

	1 angular mil	$= 10^{-3}$ radians (rad)
10 angular mils	= 1 centrad	$= 10^{-2}$ rad
100 centrads	= 1 radian	

	1 second	$= \frac{1}{3600}°$
60 seconds	= 1 minute	$= \frac{1}{60}°$
60 minutes	= 1 degree	
90 degrees	= 1 right angle	
4 right angles	= 1 circle*	$= 360°$

30 degrees	= 1 sign
45 degrees	= 1 octant
60 degrees	= 1 sextant

	1 centesimal second	$= 9 \times 10^{-5}°$
100 centesimal seconds	= 1 centesimal minute	$= 9 \times 10^{-3}°$
100 centesimal minutes	= 1 grade	$= 9 \times 10^{-1}°$
100 grades	= 1 right angle	

* = turn
NB: The recommended unit of angle is the radian or, alternatively, the degree.

3 Angular velocity and acceleration

Table 31: Common units of angular velocity

	degrees per second °/s	radians per second rad/s	(full) turns per second turn/s
1 °/s =	1	$\dfrac{\pi}{180}$	$\dfrac{1}{360}$
1 rad/s =	$\dfrac{180}{\pi}$	1	$\dfrac{1}{2\pi}$
1 turn/s =	360	2π	1

(1) $\pi = 3 \cdot 141\ 59$ to six significant figures.

(2) The common units of angular acceleration are the degree per second squared, the radian per second squared and the (full) turn per second squared: the conversion factors are the same as those given in the table.

4 Area

Table 32: Metric units of area

1 shed $= 10^{-52}$ square metres (m²)
1 barn $= 10^{-28}$ m²
1 are $= 10^2$ m²

Table 33: Imperial units of area

		1 (square) rod* $=$	$30\frac{1}{4}$ square yards (yd²)
16	(square) rods $=$ 1 (square) chain $=$	484	yd²
$2\frac{1}{2}$	(square) chains $=$ 1 rood $=$	1 210	yd²
4	roods $=$ 1 acre $=$	4 840	yd²

$\dfrac{\pi}{4} \times 10^{-6}$ square inches $= 1$ circular mil

$\dfrac{\pi}{4}$ square inches $= 1$ circular inch

* $=$ (square) pole $=$ (square) perch

5 Energy

Entries in part 1 (Units) give definitions of the wide variety of energy units listed here. The units cannot be easily tabulated.

atomic unit of energy	kilogrammetre
British thermal unit	kilowatt-hour
calorie	litre atmosphere
Centigrade heat unit	micri-erg
duty	quantum
electronvolt	Q-unit
erg	therm
frigorie	thermie
joule*	thousandth mass unit

* Recommended unit

6 Force

Table 34: Metric and metric-derived units of force

1 dyne	$= 10^{-5}$	newton (N)
1 crinal	$= 10^{-1}$	N
1 sthène	$= 10^3$	N

1 gramme-force	$=$ 0·009 806 65 N
1 kilogramme-force	$=$ 9·806 65 N

NB: Imperial units of force include the ouncedal, the poundal, the pound-force and the tondal.

Appendix 5

7 Length

Table 35: Metric units of length

1 fermi	$= 10^{-15}$	metre (m)
1 stigma	$= 10^{-12}$	m
1 ångström	$= 10^{-10}$	m
1 micron	$= 10^{-6}$	m
1 metric line	$= 10^{-3}$	m
1 centimetre	$= 10^{-2}$	m
1 kilometre	$= 10^{3}$	m
1 quadrant	$= 10^{7}$	m
1 spat	$= 10^{12}$	m

Table 36: Imperial units of length

		1 douzième	$= \frac{1}{1728}$	foot (ft)
12	douzièmes	= 1 line	$= \frac{1}{144}$	ft
4	lines	= 1 barleycorn	$= \frac{1}{36}$	ft
3	barleycorns	= 1 inch	$= \frac{1}{12}$	ft
12	inches	= 1 foot		
3	feet	= 1 yard		
$5\frac{1}{2}$	yards	= 1 rod*	$=$	$16\frac{1}{2}$ ft
4	rods	= 1 chain	$=$	66 ft
10	chains	= 1 furlong	$=$	660 ft
8	furlongs	= 1 mile	$=$	5 280 ft
3	miles	= 1 league	$=$	15 840 ft

0·001	inch	= 1 mil	$= \dfrac{1}{12\,000}$	ft
$\frac{3}{4}$	inch	= 1 digit	$= \frac{1}{16}$	ft
3	inches	= 1 palm	$= \frac{1}{4}$	ft
4	inches	= 1 hand	$= \frac{1}{3}$	ft
9	inches	= 1 span	$= \frac{3}{4}$	ft

$2\frac{1}{2}$	feet	= 1 military pace		
$4\frac{1}{2}$	feet	= 1 goad		
5	feet	= 1 pace		
100	links	= 1 chain	$=$	66 ft
		1 engineer's chain	$=$	100 ft

6	feet	= 1 fathom		
$2\frac{1}{2}$	fathoms	= 1 nautical chain	$=$	15 ft
400	nautical chains	= 1 sea mile	$=$	6 000 ft

* = pole = perch

8 Luminance

All the units of luminance are given in table 8, with the exception of the skot:

$$1 \text{ skot} = 0 \cdot 001 \text{ apostilb.}$$

9 Mass

Table 37: Metric and metric-derived units of mass

1 gamma	=	10^{-9} kilogramme (kg)
1 gramme	=	10^{-3} kg
1 quintal	=	10^{2} kg
1 tonne	=	10^{3} kg

1 point	$= 2 \times 10^{-6}$ kg
1 metric grain	$= 5 \times 10^{-5}$ kg
1 metric carat	$= 2 \times 10^{-4}$ kg
1 mounce	$= 2 \cdot 5 \times 10^{-2}$ kg
1 livre	$= 0 \cdot 5$ kg

1 glug	$= 0 \cdot 980\ 665$ kg
1 metric slug	$= 9 \cdot 806\ 65$ kg

$$1 \text{ UK assay ton} = \frac{49}{1\ 500} \text{ kg}$$

$$1 \text{ US assay ton} = \frac{7}{240} \text{ kg}$$

Table 38: Imperial units of mass (avoirdupois) used in the UK

	1 dram	$= \frac{1}{256}$	pound (lb)
16 drams	= 1 ounce	$= \frac{1}{16}$	lb
16 ounces	= 1 pound		
14 pounds	= 1 stone		
2 stones	= 1 quarter*	=	28 lb
4 quarters	= 1 hundredweight	=	112 lb
20 hundredweights	= 1 ton		= 2 240 lb

7 000 grains	= 1 pound
100 pounds	= 1 cental
1 000 pounds	= 1 kip

1 slug	$\approx$	32·174 0 lb

* The quarter has also been used for $\frac{1}{4}$ stone ($3\frac{1}{2}$ pounds).
NB: The grain is equal in size to the apothecaries' and troy grains.

Table 39: Imperial units of mass (avoirdupois); US variations

100 pounds	= 1 (short) hundredweight		
5 (short) hundredweights	= 1 quarter	=	500 lb
4 quarters	= 1 (short) ton		= 2 000 lb

112 pounds	= 1 long hundredweight		
20 long hundredweights	= 1 long ton*		= 2 240 lb

* = gross ton

Appendix 5

Table 40: Imperial units of mass (apothecaries')

20 grains (gr)	= 1 scruple		
3 scruples	= 1 drachm*	=	60 gr
8 drachms	= 1 ounce	=	480 gr
12 ounces	= 1 pound	=	5 760 gr

* In the US the unit is spelled dram.
NB: The grain is equal in size to the avoirdupois and troy grains.

Table 41: Imperial units of mass (troy)

4 grains (gr)	= 1 carat		
6 carats	= 1 pennyweight	=	24 gr
20 pennyweights	= 1 ounce	=	480 gr
12 ounces	= 1 pound (lb tr)	=	5 760 gr
$2\frac{1}{2}$ pounds	= 1 mast		
10 masts	= 1 quarter	=	25 lb tr
4 quarters	= 1 hundredweight	=	100 lb tr
20 hundredweights	= 1 ton		= 2 000 lb tr

NB: The grain is equal in size to the avoirdupois and apothecaries' grains.

10 Pitch interval

Table 42: Units of pitch interval

		cents	octaves	savarts	modified savarts
1 cent	=	1	$\dfrac{1}{1\,200}$	$\dfrac{5}{6}\log 2$	0·25
1 octave	=	1 200	1	1 000 log 2	300
1 savart	=	$\dfrac{1\cdot2}{\log 2}$	$\dfrac{10^{-3}}{\log 2}$	1	$\dfrac{0\cdot3}{\log 2}$
1 modified savart	=	4	$\dfrac{1}{300}$	$3\frac{1}{3}\log 2$	1

(1) log 2 = 0·301 030 to six significant figures.
(2) The recommended unit of pitch interval is the cent.
(3) The conversion factors in the table arise from the formula which gives the pitch interval I between two frequencies f_1, f_2:

$$I = k \log\frac{f_2}{f_1}.$$

For I in cents	$k = 1\,200/\log 2$
octaves	$1/\log 2$
savarts	10^3
modified savarts	$300/\log 2$

11 Pressure

Table 43: Metric and metric-derived units of pressure

1 barye	=	10^{-1}	pascal (Pa)
1 vac	=	10^2	Pa
1 pièze	=	10^3	Pa
1 bar	=	10^5	Pa
1 torr	$= \dfrac{101\ 325}{760}$		Pa
1 atmosphere	= 101 325		Pa
1 millimetre of water	=	9·806 65	Pa
1 technical atmosphere	=	98 066·5	Pa
1 millimetre of mercury =		133·322 387 415	Pa
1 centimetre of mercury =		1 333·223 874 15	Pa

NB: Imperial units of pressure include the psi and the tsi.

12 Radioactivity measurements

Entries in part 1 (Units) give definitions of the wide variety of radioactivity units listed here. The units cannot be easily tabulated.

absorbed ionising radiation dose
    (due to corpuscles and radiation)   gramme-rad (integral)
        rad   (inorganic)
        rem   (organic)
    (due to corpuscles)   rep
absorbed radioactive energy   gramme-röntgen
neutron flux   chad
radiation dose
    (due to corpuscles and radiation)   energy unit
        Pastille dose
    (due to γ- and X-rays)   röntgen
        skin erythema dose
    (due to γ-rays)   milligramme-hour
        sievert
    (due to X-rays)   D-unit
        e-unit
        milliampère-second
        millicurie-destroyed
    (due to neutrons)   neutron röntgen
radiation dose rate = intensity
    (due to γ-rays)
    (due to X-rays)   röntgen-per-hour-at-one-metre
        E-unit
        R-unit (Solomon)
        R-unit (German)

radioactive concentration

eman
mache

radioactive disintegration rate
    = activity

curie
rutherford

radioactive power

megaelectronvolt curie

reactivity

cent
dollar
inhour
k
nile

relative concentration

strontium unit

13 Temperature

Table 44: Units of temperature value

	degrees Celsius °C	degrees Fahrenheit °F	kelvins* K	degrees Rankine °R	degrees Réaumur °r
x °C =	x	$\frac{9}{5}x + 32$	$x + 273\cdot15$	$\frac{9}{5}x + 491\cdot67$	$\frac{4}{5}x$
x °F =	$\frac{5}{9}(x - 32)$	x	$\frac{5}{9}(x + 459\cdot67)$	$x + 459\cdot67$	$\frac{4}{9}(x - 32)$
x K =	$x - 273\cdot15$	$\frac{9}{5}x - 459\cdot67$	x	$\frac{9}{5}x$	$\frac{4}{5}(x - 273\cdot15)$
x °R =	$\frac{5}{9}(x - 491\cdot67)$	$x - 459\cdot67$	$\frac{5}{9}x$	x	$\frac{4}{9}(x - 491\cdot67)$
x °r =	$\frac{5}{4}x$	$\frac{9}{4}x + 32$	$\frac{5}{4}x + 273\cdot15$	$\frac{9}{4}x + 491\cdot67$	x

* = degrees Kelvin (°K).
NB: The recommended unit of temperature value is the kelvin or, alternatively, the degree Celsius.

Table 45: Units of temperature interval (or difference)

	Celsius degrees* degC	kelvin† K	Fahrenheit degrees degF	Rankine degrees degR	Réaumur degrees deg r
1 degC = 1 K =	1		$\frac{9}{5}$		$\frac{4}{5}$
1 degF = 1 degR =	$\frac{5}{9}$		1		$\frac{4}{9}$
1 deg r =	$\frac{5}{4}$		$\frac{9}{4}$		1

* = degrees (deg); † = Kelvin degrees (degK).

NB: The recommended unit of temperature interval is the kelvin.

14 Time

Table 46 Units of time

60 seconds	= 1 minute		
60 minutes	= 1 hour	=	3 600 seconds (s)
24 hours	= 1 day	=	86 400 s
7 days	= 1 week	=	604 800 s
4 weeks	= 1 lunar month	=	2 419 200 s

1 blink	= 10^{-5} day
1 cé*	= 10^{-2} day
1 cron	= 10^{6} years

* = degree

Table 47: The number of days in each month

days	seconds	months
31	2 678 400	January, March, May, July, August, October, December
30	2 592 000	April, June, September, November
29	2 505 600	February (in a leap year)
28	2 419 200	February (in an ordinary year)

NB: The lengths of the various kinds of year are tabulated under the entry 'year' in part 1 (Units).

15 Volume and capacity

Table 48: Metric units of volume

1 lambda	= 10^{-9}	cubic metres (m³)
1 mil	= 10^{-6}	m³
1 litre	= 10^{-3}	m³
1 stère	= 1	m³

Table 49: Imperial units of capacity used in the UK

	1 minim	= $\frac{1}{9\,600}$ pint (pt)	
60 minims	= 1 fluid drachm	= $\frac{1}{160}$ pt	
8 fluid drachms	= 1 fluid ounce	= $\frac{1}{20}$ pt	
5 fluid ounces	= 1 gill	= $\frac{1}{4}$ pt	
4 gills	= 1 pint		
2 pints	= 1 quart		
2 quarts	= 1 pottle	=	4 pt
2 pottles	= 1 gallon	=	8 pt
2 gallons	= 1 peck	=	16 pt
4 pecks	= 1 bushel	=	64 pt
4 bushels	= 1 coom(b)*	=	256 pt
2 coombs	= 1 quarter	=	512 pt
4½ quarters	= 1 chaldron	=	2 304 pt
5 quarters	= 1 wey	=	2 560 pt

* Also spelled comb

Table 50: Imperial units of liquid capacity used in the US

$$
\begin{array}{lll}
& 1 \text{ minim} & = \tfrac{1}{7\,680} \text{ liquid pint (liq pt)} \\
60 \text{ minims} & = 1 \text{ fluid dram} & = \tfrac{1}{128} \text{ liq pt} \\
8 \text{ fluid drams} & = 1 \text{ fluid ounce} & = \tfrac{1}{16} \text{ liq pt} \\
4 \text{ fluid ounces} & = 1 \text{ gill} & = \tfrac{1}{4} \text{ liq pt} \\
4 \text{ gills} & = 1 \text{ liquid pint} \\
2 \text{ liquid pints} & = 1 \text{ liquid quart} \\
4 \text{ liquid quarts} & = 1 \text{ gallon} & = \quad 8 \text{ liq pt} \\
42 \text{ gallons} & = 1 \text{ barrel} & = \quad 336 \text{ liq pt}
\end{array}
$$

Table 51: Imperial units of dry capacity used in the US

$$
\begin{array}{lll}
2 \text{ dry pints (dry pt)} & = 1 \text{ dry quart} \\
8 \text{ dry quarts} & = 1 \text{ peck} & = \quad 16 \text{ dry pt} \\
4 \text{ pecks} & = 1 \text{ bushel} & = \quad 64 \text{ dry pt} \\
36 \text{ bushels} & = 1 \text{ chaldron} & = 1\,024 \text{ dry pt}
\end{array}
$$

16 Yarn counts

The units quoted in table 52 are those of line density and reciprocal line density.

Table 52: Metric and imperial yarn counts

$$
\begin{array}{lllll}
\text{direct} & 1 \text{ denier} & = 1 \text{ gramme per 9 000 metres} & = \tfrac{1}{9} & \text{tex} \\
& 1 \text{ drex} & = 1 \text{ gramme per ten kilometres} & = \tfrac{1}{10} & \text{tex} \\
& 1 \text{ poumar} & = 1 \text{ pound per million yards} & \approx 0{\cdot}496\,055 & \text{tex} \\
& 1 \text{ tex} & = 1 \text{ gramme per kilometre}
\end{array}
$$

$$
\begin{array}{llll}
\text{indirect} & 1 \text{ typp} & = 1\,000 \text{ yards per pound} & \approx 0{\cdot}002\,015\,91/\text{tex}
\end{array}
$$

Appendix 6

ARBITRARY SCALES OF MEASUREMENT

This appendix contains details of quantities that are quoted by means of a reference number or letter rather than by actual size. The reference number can represent a particular value (eg wire gauge); it can represent a range of values (eg the Beaufort scale); it can also represent an arbitrarily-designated size that has no quantitative equivalent (eg Moh's scale).

Table 53: The Beaufort scale of wind speeds

| Beaufort number | descriptive title | land specification | range of wind speed | |
			mi/h	km/h
0	calm	smoke rises vertically	less than 1	less than 1
1	light air	direction shown by smoke but not wind vane	1 to 3	1 to 5
2	light breeze	wind felt on face; leaves rustle; vane moved by wind	4 to 7	6 to 12
3	gentle breeze	leaves and small twigs in constant motion; wind extends light flag	8 to 12	13 to 20
4	moderate (breeze)	raises dust and loose paper; small branches moved	13 to 18	21 to 29
5	fresh (breeze)	small trees in leaf begin to sway; crested wavelets on inland waters	19 to 24	30 to 39
6	strong (breeze)	large branches in motion; whistling in telegraph wires; umbrellas used with difficulty	25 to 31	40 to 50
7	moderate gale	whole trees in motion; inconvenience felt when walking against wind	32 to 38	51 to 61
8	fresh gale	breaks twigs off trees; generally impedes progress	39 to 46	62 to 74
9	strong gale	slight structural damage (chimney pots and slates removed)	47 to 54	75 to 87
10	whole gale	seldom experienced inland; trees uprooted; considerable structural damage	55 to 63	88 to 102
11	storm	very rarely experienced; widespread damage	64 to 75	103 to 120
12	hurricane	most violent	above 75	above 120

Braces in the descriptive title column group numbers 1–3 as "wind light" and numbers 7–8 as "gale".

This scale is employed by meteorologists. The Beaufort numbers N are related to the average wind speed v by the empirical formula

$$v = kN^{3/2},$$

where
$$k = 1 \cdot 87 \text{ mi/h}$$
$$= 3 \cdot 0 \text{ km/h}.$$

Table 54: The Mercalli scale of earthquake intensities (modified form)

scale number	descriptive title	characteristics	maximum ground acceleration m/s^2	magnitude of highest intensity (Richter scale)
I	instrumental	detected only by seismographs	0·01	< 3·5
II	feeble	noticed only by sensitive people	0·025	3·5
III	slight	like vibrations due to passing lorry; felt by people at rest, especially on upper floors	0·05	4·2
IV	moderate	felt by people when walking; loose objects and stationary vehicles rock	0·1	4 5
V	rather strong	generally felt; sleeping people awakened	0·25	4·8
VI	strong	trees sway, suspended objects swing; damage by movement of loose objects	0·5	5·4
VII	very strong	general alarm; walls crack and plaster falls	1	6·1
VIII	destructive	masonry fissures; chimneys fall; poorly constructed buildings damaged	2·5	6·5
IX	ruinous	some houses collapse where ground begins to crack; pipes break open	5	6·9
X	disastrous	ground cracks badly; many buildings destroyed; railway lines bent; landslides on steep slopes	7·5	7·3
XI	very disastrous	few buildings remain standing; bridges destroyed; all services out of action; great landslides and floods	9·8	8·1
XII	catastrophic	total destruction; objects thrown into the air; ground rises and falls in waves	> 9·8	> 8·1

This scale is employed by geologists and geophysicists.

Table 55: Moh's scale of relative hardnesses

original form		modified form	
scale number	mineral type	scale number	mineral type
1	talc	1	talc
2	rock salt, gypsum	2	rock salt, gypsum
3	calcite	3	calcite
4	fluorite	4	fluorite
5	apatite	5	apatite
6	feldspar	6	feldspar
		7	vitreous fused silica
7	quartz	8	quartz
		9	garnet
8	topaz	10	topaz
		11	fused zirconia
9	corundum		
		12	fused aluminia
		13	silicon carbide
		14	boron carbide
10	diamond	15	diamond

Rocks and minerals are given the reference number typified by the example quoted against a given number; half-way values (eg $4\frac{1}{2}$) are also used. Each mineral can be scratched by those higher up in the series, and will scratch those lower down in the series.

Table 56: Wire diameter and sheet-metal thickness gauges

name	abbreviation	range of gauge numbers	size range		notes
American steel and music wire gage	MWG	6/0 to 45	0·004 0 in	to 0·180 0 in	1
American wire gage	AWG	40 to 4/0	0·003 145 in	to 0·460 0 in	1
American zinc gage	—	1 to 25	0·002 0 in	to 0·250 0 in	
Birmingham gauge	BG	52 to 15/0	0·000 95 in	to 1·000 in	2
Birmingham wire gauge	BWG	36 to 4/0	0·004 0 in	to 0·454 0 in	
Brown and Sharpe gage	B & S	40 to 4/0	0·003 145 in	to 0·460 0 in	1
English music wire gauge	—	2 to 30	0·010 5 in	to 0·075 0 in	
English zinc gauge	ZG	1 to 21	0·004 0 in	to 0·070 0 in	
German sheet gauge	—	32 to 1	0·18 mm	to 5·50 mm	
ISO metric preferred series	R40, R20, R10	—	0·020 mm	to 25·0 mm	3
Lancashire pinion wire gauge	LPG	80 to H1	0·013 in	to 0·494 in	
New Westphalian music wire gauge	—	0 to 30	0·125 mm	to 2·000 mm	
Standard wire gauge	SWG	50 to 7/0	0·001 0 in	to 0·500 in	
Stubbs wire gauge	—	80 to J	0·013 in	to 0·277 in	2
US manufacturers' standard gage	MSG	39 to 7/0	0·005 6 in	to 0·478 3 in	4
US standard gage	USG	39 to 7/0	0·005 9 in	to 0·500 0 in	
US steel wire gage	USSWG	50 to 0	0·004 4 in	to 0·306 5 in	

In general gauge numbers increase or decrease in 1s; through zero they run as . . 2, 1, 0, 2/0, 3/0. . . . Where letters are introduced, the series runs
. . 2, 1, A, B, . . . Y, Z, A1, B1. . . .
1 The AWG and B & S gage are identical.
2 Legal standard in the UK up to 31 January 1964.
3 International standard and only legal standard in the UK since 1 February 1964 (by 1963 Weights and Measures Act). The R20 series contains
every alternate value of the R40 series; the R10 series contains every alternate value of the R20 series.
4 Identical to the LPG up to gauge letter J.

Appendix 7

MUSICAL INTERVALS

Table 57 lists the more important intervals that are utilised in the structure of the modern Western musical scales. The size of each interval is given, both in cents and as a decimal, to six significant figures. Values in other units can be obtained from the value in cents by multiplying by the following factors:

$$\frac{1}{1200} \approx 0{\cdot}000\ 833\ 333 \quad \text{(interval in octaves)}$$
$$\frac{1}{12} \log 2 \approx 0{\cdot}250\ 858 \quad \text{(interval in savarts)}$$
$$\frac{1}{4} = 0{\cdot}25 \quad \text{(interval in modified savarts)}$$

The abbreviations used in the last column are:

      E   interval on equally tempered scale
      J   interval on justly (ie evenly) tempered scale

Table 57: Musical intervals in order of ascending size

size of interval			name of interval	
cents	decimal	fraction		
0	1·0	1:1	E, J	unison
1	000 578	$(2^{1/1\,200})$	E, J	cent
3·986 31	002 305	$(2^{1/1\,000\,\log 2})$	E, J	savart
4	002 313	$(2^{1/300})$	E, J	modified savart
21·506 3	012 5	81:80	J	(syntonic) comma
50	029 303	$(2^{1/24})$	E	quartertone
70·672 4	041 667	25:24	J	smaller chromatic semitone
92·178 7	054 688	135:128	J	larger chromatic semitone
100	059 463	$(2^{1/12})$	E	semitone
111·731	066 667	16:15	J	minor 2nd = diatonic semitone
182·404	111 111	10:9	J	lesser 2nd = lesser (or minor) whole tone
200	122 462	$(2^{1/6})$	E	2nd = (major or minor) tone
203·910	125	9:8	J	major (larger) 2nd (or, whole tone)
223·463	137 777	256:225	J	diminished minor 3rd
274·582	171 875	75:64	J	augmented 2nd
300	189 207	$(2^{1/4})$	E	augmented 2nd = minor 3rd
315·641	2	6:5	J	minor 3rd
386·314	25	5:4	J	major 3rd
400	259 921	$(2^{1/3})$	E	major 3rd = diminished 4th
427·372	28	32:25	J	diminished 4th
478·492	318 359	675:512	J	augmented 3rd
498·045	333 333	4:3	J	perfect 4th
500	334 840	$(2^{5/12})$	E	augmented 3rd = perfect 4th
590·224	406 25	45:32	J	augmented 4th
600	414 214	$(2^{1/2})$	E	augmented 4th = diminished 5th
609·776	422 222	64:45	J	diminished 5th
700	498 307	$(2^{7/12})$	E	perfect 5th
701·955	5	3:2	J	perfect 5th
772·627	562 5	25:16	J	augmented 5th
800	587 401	$(2^{2/3})$	E	augmented 5th = minor 6th
813·686	6	8:5	J	minor 6th
884·359	666 667	5:3	J	major 6th
900	681 793	$(2^{3/4})$	E	major 6th
925·418	706 667	128:75	J	diminished 7th
976·537	757 812	225:128	J	augmented 6th
996·090	777 778	16:9	J	minor 7th
1 000	781 797	$(2^{5/6})$	E	augmented 6th = minor 7th
1 088·27	875	15:8	J	major 7th
1 100	887 749	$(2^{11/12})$	E	major 7th = diminished 8th
1 107·82	896 296	256:135	J	diminished 8th
1 180·45	1·997 070	2 025:1 024	J	augmented 7th
1 200	2	2:1	{ J	octave
			{ E	augmented 7th = octave

Appendix 8

SCALE RATIOS FOR MAPS

Table 58 gives the international scale ratios recommended for use on maps, plans and diagrams. It is envisaged that these will soon replace the presently used 'imperial' scale ratios.

In the second column the following letters are used:

A	maps	E	sketch schemes, etc
B	town surveys	F	location drawings
C	surveys and layouts	G	component and assembly detail
D	site and key plans		drawings

The scale ratios marked with an asterisk are temporarily included during the changeover from 'imperial' to international scale ratios.

Table 58: Scale ratios

scale	use	nearest 'imperial' equivalent		
1:1 000 000	A	1:1 000 000	$\frac{1}{16}$ in to	1 mi approximately
1: 500 000	A	1: 625 000	$\frac{1}{10}$ in to	1 mi approximately
1: 200 000	A	1: 250 000	$\frac{1}{4}$ in to	1 mi approximately
1: 100 000	A	1: 126 720	$\frac{1}{2}$ in to	1 mi
1: 50 000	AB	1: 63 360	1 in to	1 mi
1: 20 000	B	1: 25 000	$2\frac{1}{2}$ in to	1 mi approximately
1: 10 000	B	1: 10 560	6 in to	1 mi
1: 5 000	B	—		—
*1: 2 500	BC	} 1: 2 500	1 in to 200 ft	approximately
1: 2 000	C			
*1: 1 250	CD	} 1: 1 250	1 in to 100 ft	approximately
1: 1 000	CD			
1: 500	CD	1: 500	1 in to 40 ft	approximately
—	—	1: 384	$\frac{1}{32}$ in to	1 ft
1: 200	EF	1: 192	$\frac{1}{16}$ in to	1 ft
1: 100	EF	1: 96	$\frac{1}{8}$ in to	1 ft
1: 50	F	1: 48	$\frac{1}{4}$ in to	1 ft
1: 20	G	1: 24	$\frac{1}{2}$ in to	1 ft
1: 10	G	1: 12	1 in to	1 ft
1: 5	G	1: 4	3 in to	1 ft
1: 1	G	1: 1	1 in to	1 in

Appendix 9

UNIVERSAL PHYSICAL CONSTANTS

Most standard reference works provide a more or less complete list of the values of universal constants. The constants given in table 59 have been limited to those directly related to the units and quantities included in this work.

Table 59: Universal physical constants

quantity	Symbol[1]	size in SI units[2]		notes on size[3]	notes
standard acceleration of free fall	g_n	9·806 65	m/s²	(E)	4
standard atmospheric pressure	p	1·013 25 $\times 10^5$	Pa	(E)	
velocity of propagation of electromagnetic waves in vacuo	c_o	(2·997 925 0 ± 0·000 001 0) $\times 10^8$	m/s	(M)	
Planck's constant	h	(6·624 9 ± 0·000 3) $\times 10^{-34}$	J s	(M)	
	$\hbar$	(1·054 39 ± 0·000 05) $\times 10^{-34}$	J s	$= h/2\pi$ (E)	5
magnetic constant	μ_o	1·256 64 $\times 10^{-6}$	H/m	$= 4\pi \times 10^{-7}$ H/m (E)	6
electric constant	ϵ_o	8·854 16 $\times 10^{-12}$	F/m	$= 10^7/4\pi\, c_o{}^2$ F/m (E)	7
charge on the electron	e	(1·602 10 ± 0·000 07) $\times 10^{-19}$	C	(M)	8
mass of the electron	m	(9·108 4 ± 0·000 3) $\times 10^{-31}$	kg	(M)	

1 International.
2 The unit symbols employed are:

C coulomb   J joule   N newton
F farad   kg kilogramme   Pa pascal
H henry   m metre   s second

3 (E) = exact, by definition
  (M) = measured value
4 Defined by the 3rd Conférence Générale des Poids et Mesures of 1901.
5 The symbol is pronounced 'h-crossed'.
6 Formerly called the permeability of free space. The unrationalised size $\overset{*}{\mu}_o$ is given by $\overset{*}{\mu}_o = \mu_o/4\pi$.
7 Formerly called the permittivity of free space. The unrationalised size $\overset{*}{\epsilon}_o$ is given by $\overset{*}{\epsilon}_o = 4\pi\,\epsilon_o$.
8 Also called the elementary charge.

Appendix 10

THE GREEK ALPHABET

Table 60: The Greek alphabet

capital	lowercase	name	capital	lowercase	name
A	α	alpha	N	ν	nu
B	β	beta	Ξ	ξ	xi
Γ	γ	gamma	O	ο	omicron
Δ	δ	delta	Π	π	pi
E	ε	epsilon	P	ρ	rho
Z	ζ	zeta	Σ	σ	sigma
H	η	eta	T	τ	tau
Θ	θ	theta	Υ	υ	upsilon
I	ι	iota	Φ	φ	phi
K	κ	kappa	X	χ	chi
Λ	λ	lambda	Ψ	ψ	psi
M	μ	mu	Ω	ω	omega

Note: There also exist a final form of lowercase sigma *s*, and a form of lowercase pi known as Dorian pi (curly pi) ϖ.

Appendix 11

UNIT SYMBOLS: AN ALPHABETICAL LIST

Unit symbols may be classified as follows:

1 Recommended
2 One of two (or more) equally acceptable symbols
3 In use, but deprecated; an alternative is preferred
4 No longer in use

In the following list no indication is given of the class. For this information reference should be made to the appropriate entry in part 1 (Units).

In general, the abbreviations of metric multiples and submultiples are not given, nor are those of combination units (eg metre per second).

Unit symbols with lowercase initial letters are entered separately from similar ones with capital initial letters, and immediately precede them. Greek alphabet entries are listed after the Roman alphabet entries; the list concludes with non-alphabet unit symbols.

a	are, year	$Btu_{39/40}$	thirty-nine degrees Fahrenheit British thermal unit
A	ampère, ampère-turn, ångström, A-size	$Btu_{60/61}$	sixty degrees Fahrenheit British thermal unit
Å	ångström	bu	bushel
°A	degree absolute		
A_{abs}	absolute ampère		
A_{int}	international ampère	c	curie, cycle
abA	abampère	C	coulomb, C-size
abC	abcoulomb	^{c}	centesimal minute
abF	abfarad	C_{abs}	absolute coulomb
abH	abhenry	C_{int}	international coulomb
abS	absiemens	°C	degree Celsius
abT	abtesla	cal	(gramme-)calorie
abV	abvolt	Cal	large calorie
abWb	abweber	cal_{IT}	international table calorie
abΩ	abohm	cal_{mean}	mean calorie
ACI	acoustical comfort index	cal_{tc}	thermochemical calorie
amp	ampère	cal_4	four degrees Celsius calorie
amu	atomic mass unit	cal_{15}	fifteen degrees Celsius calorie
asb	apostilb		
at	technical atmosphere	^{cc}	centesimal second
At	ampère-turn	cd	candela, new candle
AT	ampère-turn	ch	chain, cheval vapeur
atm	atmosphere	CHU	Centigrade heat unit
atmo-m	atmo-metre	CHU_{mean}	mean Centigrade heat unit
au	astronomical unit	Ci	curie
AU	ångström (unit), astronomical unit	cm	centimetre
awu	atomic weight unit	cm-c	centimetre candle
		cmHg	(conventional) centimetre of mercury
b	bar, barn, bel	cps	cycle per second
B	bel, brewster, B-size	c/s	cycle per second
°B	degree Baumé	ctl	cental
bbl	dry barrel	cwt	hundredweight
bd	bougie décimale	cwt tr	troy hundredweight
bd ft	board foot		
Bi	biot	d	day
BThU	British thermal unit	D	debye
Btu	British thermal unit	db	decibel
BTU	British thermal unit	dB	decibel
Btu_{IT}	international table British thermal unit	deg	Celsius degree, degree (of arc), degree (of temperature), kelvin
Btu_{mean}	mean British thermal unit		

degA	absolute degree	Gs	gauss
degC	Celsius degree	gwt	gramme-weight
degF	Fahrenheit degree		
degK	kelvin, Kelvin degree	h	hour
deg r	Réaumur degree	H	henry
degR	Rankine degree	H_{abs}	absolute henry
dr	dram	H_{int}	international henry
dram ap	apothecaries' dram	HB	Brinell hardness number
dwt	pennyweight	hhd	hogshead
dyn	dyne	HK	Hefnerkerze
		hp	horsepower
E	eötvös, erlang	HR	Rockwell hardness number
EBR	equivalent biological rönt-	HV	Vickers hardness number
	gen	Hz	hertz
eV	electronvolt		
		IÅ	international ångström
f	fors	ih	inhour
F	farad, fraunhofer	Imch	millicurie-of-intensity-hour
°F	degree Fahrenheit	in	inch
F_{abs}	absolute farad		
F_{int}	international farad	J	joule
fc	foot-candle	J_{abs}	absolute joule
fg	frigorie	J_{int}	international joule
fl dr	fluid drachm, fluid dram		
fl oz	fluid ounce	K	kayser, kelvin
Fr	franklin	°K	degree Kelvin, kelvin
ft	foot	kcal	kilocalorie
ft-L	foot-lambert	kg	kilogramme
fu	flux unit	kg-eq	kilogramme-equivalent
FU	Finsen unit	kgf	kilogramme-force
fur	furlong	kgwt	kilogramme-weight
		kn	knot
g	gramme	kp	Kilopond
G	gauss, grav	kWh	kilowatt-hour
g	grade		
gal	gallon	l	litre
Gal	galileo	L	lambert
Gb	gilbert	l atm	litre-atmosphere
g-eq	gramme-equivalent	lb	pound (mass, avoirdupois)
gf	gramme-force	lb	troy pound
gi	gill	Lb	Pound(-force)
Gi	gilbert	lb ap	apothecaries' pound
gmol	gramme-molecule	lb apoth	apothecaries' pound
gr	grain	lbf	pound-force

lb t	troy pound	P	poise
lb tr	troy pound	Pa	pascal
lbwt	pound-weight	pc	parsec
li	link	pdl	poundal
lm	lumen	pk	peck
LU	loudness unit	Pl	poiseuille
lx	lux	PNdB	perceived noise decibel
ly	light year	praGb	pragilbert
		praGs	pragauss
		praMx	pramaxwell
m	metre, minute (of time)	praOe	praoersted
M	Richter magnitude	PS	Pferdestärke
mag	magnitude (stellar)	pt	pint
mA s	milliampère-second	pz	pièze
m-atm	metre-atmosphere		
mc	metre-candle		
mcd	millicurie-destroyed	q	quintal
mCi h	millicurie-hour	Q	Q-unit
MeV Ci	megaelectronvolt-curie	qɪ	quarter
mg h	milligramme-hour	qr tr	troy quarter
mi	mile		
min	minim, minute (of time)	r	röntgen
ml	mil	°r	degree Réaumur
mmHg	(conventional) millimetre of mercury	R	rayleigh, röntgen
		°R	degree Rankine
mmH₂O	(conventional) millimetre of water	r	radian
		RA	RA-size
mmu	millimass unit	rad	radian
mol	mole	rd	rutherford
Mx	maxwell	REM	röntgen-equivalent mammal
		REP	röntgen-equivalent physical
N	neper, newton		
n mi	nautical mile	rhm	röntgen-per-hour-at-one-metre
Np	neper		
nt	nit	Rt ∟	right angle
o	octave	s	savart, second (of time), stère
Oe	oersted		
owu	open window unit	S	siemens, spat, stokes, svedberg
oz	ounce		
oz ap	apothecaries' ounce	Sₐᵦₛ	absolute siemens
oz apoth	apothecaries' ounce	Sᵢₙₜ	international siemens
oz t	troy ounce	sb	stilb
oz tr	troy ounce	SED	skin erythema dose

sh cwt	short hundredweight	W_{int}	international watt
sh tn	short ton	Wb	weber
sh ton	short ton	Wb_{abs}	absolute weber
sn	sthène	Wb_{int}	international weber
sr	steradian		
SRA	SRA-size	X	Siegbahn unit, X-unit
st	stère	XU	Siegbahn unit, X-unit
St	stat, stokes		
statA	statampère		
statC	statcoulomb	y	year
statF	statfarad	yd	yard
statH	stathenry		
statS	statsiemens	β	Bohr magneton
statT	stattesla	γ	gamma
statV	statvolt	λ	lambda
statWb	statweber	μ	micron (length)
statΩ	statohm	μmHg	micron (pressure)
sterad	steradian	σ	stigma
SU	strontium unit, sunshine unit	Ω	ohm
		Ω_a	acoustical ohm
		Ω_{abs}	absolute ohm
t	tonne	Ω_{int}	international ohm
T	tesla, turn	Ω_m	mechanical ohm
T_{abs}	absolute tesla	Ω_s	specific acoustical ohm
T_{int}	international tesla	Ω_u	unit-area acoustical ohm
th	thermie		
TME	Technische Mass Einheit	$\mho$	mho (= siemens)
TMU	thousandth mass unit	$\mho_{abs}$	absolute mho
tonf	ton-force	$\mho_{int}$	international mho
ton tr	troy ton		
tonwt	ton-weight	$'$	foot, minute (of arc), minute (of time)
°Tw	degree Twaddell	$''$	inch, second (of arc), second (of time)
u	atomic mass unit	$°$	degree (of arc), degree (of temperature)
V	volt		
V_{abs}	absolute volt	$\times$	yard
V_{int}	international volt	$\llcorner$	right angle
VA	voltampère	$\mathfrak{Z}$	apothecaries' dram, drachm
vib	vibration		
		$\mathfrak{Z}$	apothecaries' ounce
W	watt	$\mathfrak{Э}$	scruple
W_{abs}	absolute watt	$\mathfrak{M}$	minim

Appendix 12

SYMBOLS FOR QUANTITIES: AN ALPHABETICAL LIST

The symbols may be classified as follows:

1 Recommended
2 One of two (or more) equally acceptable symbols
3 In use, but deprecated; an alternative is preferred
4 No longer in use

In the following list no indication is given of the class. For this information reference should be made to the appropriate entry in part 2 (Quantities).
Symbols consisting of a lowercase letter are entered separately from similar ones consisting of a capital letter, and immediately precede them. Greek alphabet entries are listed after the Roman alphabet entries.

a	acceleration, amplitude, attenuation coefficient, specific area, thermal diffusivity
A	area, linear current density, magnetic vector potential, sound strength (of a source), work
b	angular momentum, breadth, phase coefficient, width
B	magnetic flux density, susceptance
B_i	magnetic polarisation
c	heat capacity, molarity, phase velocity in a medium, specific heat capacity (mass basis), velocity of electromagnetic radiation in free space
c_0	velocity of electromagnetic radiation in free space
c_p	specific heat capacity (mass basis) at constant pressure
c_v	specific heat capacity (mass basis) at constant volume
C	calorific value (volume basis), capacitance, curvature, heat capacity, mechanical compliance
C_a	acoustical compliance
d	diameter, relative density, thickness
d_v	absolute humidity
D	diameter, diffusion coefficient, electric displacement
D_i	electric polarisation
e	extinction coefficient, restitution coefficient, specific energy, tensile strain
E	acoustical energy density, electric field strength, electromotive force, energy, equivalent absorption (area), exergy, illumination, internal energy, irradiance, Young's modulus of elasticity
E_e	irradiance
E_i	electrisation
E_k	kinetic energy
E_p	potential energy
E_v	illumination
f	acceleration, frequency, statical friction coefficient, stress (normal)
F	force, free energy, fuel consumption (traffic factor), luminous flux, magnetomotive force
F_m	magnetomotive force
$\mathscr{F}$	magnetomotive force
g	local acceleration of free fall
g_n	standard acceleration of free fall
G	electrical conductance, Gibbs function, rigidity modulus of elasticity, weight
h	depth, height, thermal conductance
H	enthalpy, heat release, magnetic field strength
H_i	magnetisation
H_0	horizontal component of the earth's magnetic field strength

I	electric current, enthalpy, impulse, interval (pitch), luminous intensity, moment of inertia, radiant intensity, second moment of area, sound intensity
I_a	second moment of area
I_e	radiant intensity
I_p	second polar moment of area
I_v	luminous intensity
j	magnetic dipole moment
J	current density, magnetic polarisation, moment of inertia, second polar moment of area, sound intensity
k	circular wave number, coupling coefficient (inductance), thermal conductivity
K	bulk modulus of elasticity, electric field strength, kinetic energy, photoptic total luminous efficiency, thermal conductance
K'	scotoptic total luminous efficiency
K_i	electrisation
K_m	photoptic maximum luminous efficiency
K_m'	scotoptic maximum luminous efficiency
K_λ	photoptic luminous efficiency
K_λ'	scotoptic luminous efficiency
l	length, specific latent heat
L	action, latent heat, luminance, radiance, self-inductance, sensation level, sound pressure level
L_e	radiance
L_N	equivalent loudness, power level
L_p	sensation level, sound pressure level
L_P	power level
L_v	luminance
L_W	power level
L_{12}	mutual inductance
m	electromagnetic moment, magnetic pole strength, mass, molality
m_a	acoustical mass
M	luminous emittance, magnetisation, mechanical advantage, moment, mutual inductance, radiant emittance
M_e	radiant emittance
M_v	luminous emittance
n	molar value, refractive index, rotational frequency
N	amplitude level, intensity level, loudness, moment, sound flux
p	electric dipole moment, momentum (translational), pressure, propagation coefficient
p_e	electric dipole moment
p_θ	angular momentum
P	active power, electric polarisation, equivalent loudness, force, permeance, power, radiant flux, sound flux, thermoelectric power, weight

P_q reactive power

P_s apparent power

q heat flow rate, heat flow rate density, shear stress, specific charge, specific humidity, volume flow rate

q_m mass flow rate

q_v volume flow rate

Q charge, heat energy, luminous energy, quality factor, radiant energy, reactive power

Q_e radiant energy

Q_v luminous energy

r mechanical resistance, mixing ratio, radial distance, radius, reflexion coefficient

R earth's magnetic field strength (total), electrical resistance, gravitational field strength, reluctance, sound reduction factor, thermal resistance

R_a acoustical resistance

R_m mechanical resistance, reluctance

R_s specific acoustical resistance

$\mathscr{R}$ reluctance

s displacement, distance, line segment, mechanical stiffness, path length, specific entropy

S apparent power, area, current density, entropy, loudness, Poynting vector, sensation level

S_a acoustical stiffness

$S\theta_1/\theta_2$ relative density

t customary temperature, temperature interval, thickness, time, transmissivity

T absolute temperature, kinetic energy, moment, periodic time (period), reverberation time, time constant

u initial velocity, radiant energy density, velocity component in x-direction

U electric potential difference, internal energy, magnetic potential, magnetic potential difference, potential energy, radiant energy, relative humidity, thermal conductance, volume flow rate

U_m magnetic potential difference

U_λ spectral radiant energy

$\mathscr{U}$ magnetic potential difference

v specific volume, velocity, velocity component in y-direction, volume

V electric potential, electric potential difference, potential energy, velocity ratio, vertical component of the earth's magnetic field strength, volume

V_λ photoptic luminous relative efficiency

V_λ' scotoptic luminous relative efficiency

w	energy density, mechanical impedance, velocity component in z-direction
W	radiant energy, section modulus, sound flux, specific acoustical impedance, weight, work
x	general physical variable, path length, specific humidity
X	electrical reactance, general physical variable
X_a	acoustical reactance
X_C	reactance due to capacitance
X_L	reactance due to inductance
X_m	mechanical reactance
X_s	specific acoustical reactance
Y	admittance
z	altitude
Z	acoustical impedance, electrical impedance, section modulus
Z_a	acoustical impedance
Z_m	mechanical impedance
Z_s	specific acoustical impedance
α	absorption coefficient, angle, angular acceleration, aperture conductivity, attenuation coefficient, cubical expansion coefficient, linear current density, linear expansion coefficient, temperature coefficient, thermal conductance, thermal diffusivity
α_a	absorption coefficient
α_P	Peltier coefficient
α_S	Seebeck coefficient
α_t	linear expansion coefficient
α_T	Thomson (thermoelectric) coefficient
α_V	cubical expansion coefficient
β	angle, areal expansion coefficient, cubical expansion coefficient, phase coefficient, pressure coefficient
γ	angle, cubical expansion coefficient, electrical conductivity, principal specific heat capacities ratio, propagation coefficient, shear strain, specific weight, surface tension
δ	damping coefficient, logarithmic decrement, scattering coefficient, thickness
Δ	damping coefficient
ε	absolute permittivity, tensile strain
ε_0	electric constant
ε_r	relative permittivity
η	charge density, efficiency, luminous efficiency of a source, viscosity (absolute)
θ	angle, bulk strain, customary temperature, phase, angle, temperature interval
Θ	absolute temperature
κ	bulk compressibility, coupling coefficient (inductance), magnetic

	susceptibility, principal specific heat capacities ratio, thermal diffusivity
λ	linear expansion coefficient, thermal conductivity, wavelength
λ_e	equivalent thermal conductivity
Λ	equivalent loudness, logarithmic decrement, permeance
μ	absolute permeability, attenuation coefficient, refractive index, Poisson's ratio, statical friction coefficient, viscosity (absolute)
μ'	dynamical friction coefficient
μ_0	magnetic constant
μ_r	relative permeability
μ_α	absorption coefficient
ν	frequency, kinematic viscosity, Poisson's ratio
$\bar{\nu}$	wave number
ρ	charge density, density, electrical resistivity, reflexion coefficient, thermal resistivity
ρ'	line density
ρ_m	mass resistivity
σ	electrical conductivity, leakage coefficient (inductance), stress (normal), surface density of charge, surface tension, Thomson (thermoelectric) coefficient, wave number
τ	shear stress, time constant, transmission coefficient
ϕ	angle, angular acceleration, electrical potential, fluidity, heat flow rate density, phase angle, shear strain, thermal resistivity, velocity potential
ϕ_p	relative humidity
Φ	entropy, heat flow rate, luminous flux, magnetic flux, potential energy, radiant flux, sound flux
Φ_e	radiant flux
Φ_v	luminous flux
χ	bulk compressibility, magnetic mass susceptibility
χ_e	electrical susceptibility
ψ	saturation ratio
Ψ	electric flux
ω	angular velocity, solid angle
Ω	angular velocity, gravitational potential, solid angle

Appendix 13

SCIENTISTS WHO HAVE GIVEN THEIR NAMES TO THE NAMES OF UNITS

Table 61 gives the essential biographical details of those scientists after whom units have been named. Certain people did practically all their scientific work in a country other than that of their birth. In these cases the name of the country is enclosed in brackets after the entry in the second column.

Table 61: Scientists after whom units have been named

name	country of birth	dates	unit
AMAGAT, Emile- Hilaire	France	1841–1915	Amagat density unit, Amagat volume unit
AMPÈRE, André Marie	France	1775–1836	abampère, ampère, ampère-turn, milliampère-second, statampère, thermal ampère, var, voltampère
ÅNGSTRÖM, Anders Jonas	Sweden	1814–1874	ångström, international ångström
BALMER, Johann Jakob	Switzerland	1825–1898	balmer
BAUDOT, J M E	France	1845–1903	baud
BAUMÉ, Antoine	France	1728–1804	degree Baumé
BELL, Alexander Graham	Scotland (United States)	1847–1922	bel (decibel), perceived noise decibel
BIOT, Jean Baptiste	France	1774–1862	biot
BLONDEL, A		1863–1938	blondel
BOHR, Niels Henrik David	Denmark	1885–1962	bohr magneton
BREWSTER, Sir David	Scotland	1781–1868	brewster
BRIGGS, Henry	England	1561–1630	brig
BRINELL, Johann August	Sweden	1849–1925	Brinell (hardness) number
CELSIUS, Anders	Sweden	1701–1744	degree Celsius
CHADWICK, Sir James	England	1891–	chad

name	country of birth	dates	unit
CLARK, Thomas	Scotland	1801–1867	degree Clark
CLAUSIUS, Rudolf Julius Emanuel	Germany	1822–1888	clausius
COULOMB, Charles Augustine de	France	1736–1806	abcoulomb, coulomb, statcoulomb, thermal coulomb
CURIE, Marie	Poland (France)	1867–1935	curie, megaelectronvolt-curie, millicurie-destroyed, millicurie-hour
DALTON, John	England	1766–1844	dalton
DARWIN, Charles Robert	England	1809–1882	darwin
DEBYE, Peter Joseph Wilhelm	Holland	1884–	debye
EINSTEIN, Albert	Germany (United States)	1879–1955	Einstein unit
EÖTVÖS, Baron Roland von	Hungary	1848–1919	eötvös
ERLANG, A K	Denmark	1879–1924	erlang
FAHRENHEIT, Gabriel Daniel	Germany	1686–1736	degree Fahrenheit
FARADAY, Michael	England	1791–1867	abfarad, daraf, farad, faraday, statfarad, thermal farad
FERMI, Enrico	Italy	1901–1954	fermi
FINSEN, Niels Ryberg	Denmark	1860–1904	finsen unit
FOURIER, Jean Baptiste Joseph	France	1768–1830	fourier
FRANKLIN, Benjamin	United States	1706–1790	franklin
FRAUNHOFER, Joseph von	Germany	1787-1826	fraunhofer
FRESNEL, Augustin Jean	France	1788–1827	fresnel
GALILEI, Galileo	Italy	1564–1642	Gal (galileo)
GALVANI, Luigi	Italy	1737–1798	galvat
GAUSS, Karl Friedrich	Germany	1777–1855	gauss, pragauss
GIBBS, Josiah Willard	United States	1839–1903	gibbs
GILBERT, William	England	1540–1603	gilbert, pragilbert
HARTREE, D R		1897–1958	hartree
HEFNER-ALTENECK, Friedrich von	Germany	1845–1904	Hefnerkerze
HELMHOLTZ, Hermann Ludwig Ferdinand van	Germany	1821–1894	helmholtz
HENRY, Joseph	United States	1797–1878	abhenry, henry, stathenry, thermal henry
HERSCHEL, Sir John Frederick William	England	1792–1871	herschel
HERTZ, Heinrich Rudolph	Germany	1857–1894	hertz
JOULE, James Prescott	England	1818–1889	joule
KAPP, Gisbert		1852–1922	kapp line
KAYSER, J H G		1853–1940	kayser
KELVIN, William Thomson, 1st Baron Lord	England	1824–1907	kelvin (degree Kelvin)

name	country of birth	dates	unit
LAMBERT, Johann Heinreich	Germany	1728–1777	foot-lambert, lambert
LANGLEY, Samuel Pierpont	United States	1834–1906	langley
LORENTZ, Hendrik Antoon	Holland	1853–1928	Lorentz unit
MACHE, H		1876–1954	mache
MAXWELL, James Clerk	Scotland	1831–1879	maxwell, pramaxwell
MAYER, Julius Robert von	Germany	1814–1878	mayer
McLEOD, Herbert	England	1841–1932	mcleod
MOHR, Karl Friedrich	Germany	1806–1879	Mohr cubic centimetre
MORGAN, Thomas Hunt	United States	1866–1945	morgan
NAPIER, John	Scotland	1550–1617	neper
NEWTON, Sir Isaac	England	1642–1727	newton
OERSTED, Hans Christian	Denmark	1777–1851	oersted, praoersted
OHM, Georg Simon	Germany	1787–1854	abohm, acoustical ohm, gemmho, mechanical ohm, mho mohm, specific acoustical ohm, ohm, ohma, ohmad, secohm, statohm, reciprocal ohm, thermal ohm
PARKER, H M		1910–	parker
PASCAL, Blaise	France	1623–1662	pascal
PLANCK, Max Karl Ernst	Germany	1858–1947	planck
POISEUILLE, Jean Louis Marie	France	1799–1869	poise, poiseuille
PONCELET, Jean Victor	France	1788–1867	poncelet
PREECE, Sir William Henry	Wales	1834–1913	preece
PROUT, William	Scotland	1786–1850	prout
RANKINE, William John Macquorn	Scotland	1820–1870	degree Rankine
RAYLEIGH, John William Struff, 3rd Baron Lord	England	1842–1919	ray, rayl, rayleigh
RÉAUMUR, René Antoine Ferchault de	France	1683–1757	degree Réaumur
REYNOLDS, Osborne	N Ireland	1842–1912	reyn
ROCKWELL, Stanley P	United States		Rockwell (hardness) number
RÖNTGEN, Wilhelm Konrad	Germany	1845–1923	equivalent biological röntgen, gramme-röntgen, neutron röntgen, rem, rep, röntgen, röntgen-per-hour-at-one-metre, tissue röntgen
ROWLAND, Henry Augustus	United States	1848–1901	rowland
RUTHERFORD, Ernest, 1st Baron Lord	England	1871–1937	rutherford
RYDBERG, Johannes Robert	Sweden	1854–1919	rydberg

name	country of birth	dates	unit
SABINE, Wallace Clement	United States	1868–1919	sabin
SAVART, Félix	France	1791–1841	savart, modified savart
SIEGBAHN, Karl Manne Georg	Sweden	1886–	siegbahn unit
SIEMENS, Karl Wilhelm, later Sir William	Germany (England)	1823–1883	absiemens, siemens, statsiemens
STOKES, Sir George Gabriel	England	1819–1903	stokes
STURGEON, William	England	1783–1850	sturgeon
SVEDBERG, Theoden	Sweden	1884–	svedberg
TALBOT, William Henry Fox	England	1800–1877	talbot
TESLA, Nikola	Croatia (United States)	1856–1943	abtesla, tesla, stattesla
TORRICELLI, Evangelista	Italy	1608–1647	tor, torr
TROLAND, Leonard T	United States	1889–1932	troland
VIOLLE, Jules	France	1841–1923	violle
VOLTA, Count Alessandro	Italy	1745–1827	abvolt, electronvolt, equivalent volt, megaelectronvolt-curie, statvolt, thermal volt, var, volt, voltampère
WATT, James	Scotland	1736–1819	kilowatt-hour, light-watt, watt
WEBER, Wilhelm Eduard	Germany	1804–1891	abweber, statweber, weber

Index